THE HIDDEN SEA

"With Love" - February 14, 1971

THE HIDDEN SEA

Photographs and Notes by

Douglas Faulkner

Text by C. Lavett Smith

A STUDIO BOOK   ●   THE VIKING PRESS   ●   NEW YORK

For
Albert Bronson and Yves Merlet
who loved the sea

Text Copyright 1970 in all countries of the International Copyright Union
by
C. Lavett Smith
Photographs and Notes Copyright 1970 in all countries of the International Copyright Union
by
Douglas Faulkner
All rights reserved

First published in 1970 by The Viking Press, Inc.
625 Madison Avenue, New York, N.Y. 10022

Published simultaneously in Canada by
The Macmillan Company of Canada Limited

SBN 670-37067-3
Library of Congress catalog card number: 72-117062

Printed and bound in Switzerland by C. J. Bucher Publishers, Ltd.

Acknowledgments
Harcourt, Brace & World, Inc.: "La Guerre"
from *Poems 1923-1954* by E. E. Cummings. Copyright, 1923, 1951 by E. E. Cummings. Reprinted
by permission of Harcourt, Brace & World, Inc.

CONTENTS

1. THE HIDDEN SEA

This book is dedicated to beauty in the world of nature. Grace and harmony of form and color are outstanding characteristics of natural objects, and the photographs included here amply support the thesis that of all the beauties known to man the purest are those provided by nature herself. These photographs have been carefully selected, and each shows the animal alive and in its natural setting. It is hoped that the pictures will be enjoyed for their beauty and that the text will add to that enjoyment by supplying information about the animals themselves, in the way that a libretto supplies background information about an opera, adding to our understanding and therefore to our enjoyment of the fine music.

If each of us could have accompanied the photographer on his many diving trips we would have seen more or less the same things he saw. But we certainly would not have seen them through his eyes. The photographer is our guide and our interpreter. His skills serve to bring out in precise clarity details that the ordinary observer misses or perceives only imperfectly.

Our reactions depend on many things, and our emotions are triggered by the image of the moment superimposed on a mosaic of background information assembled from many sources. Consider the crown-of-thorns starfish: to the neophyte diver, apprehensive in his new environment, the animal is a source of danger to be avoided whenever possible; to the marine biologist, its fascination lies in its relationship to other organisms, and its beauty is secondary or might be overlooked. To the photographer, the crown-of-thorns is primarily an object of beauty with its own unique perfection of geometry, texture, and color, to be enjoyed, first and foremost, for its aesthetic qualities. Through the marvels of modern photography we are able to share something of this view with the photographer, and in this one sense the photograph is actually superior to the living creature.

The only disadvantage of photographs is that, being themselves immobile, they tend to give the impression that the things they portray are also motionless. This is particularly unfortunate in the world of natural history, for living organisms are dynamic and ever-changing. Another exposure, a fraction of a second earlier or later, would not have been exactly the same. Constant change is a pervading rule in our universe.

Young individuals change continually as they breathe, feed, and grow. Eventually they wear out, die, and are replaced by their offspring. Finally, through eons of time, there are slow progressive evolutionary changes as new generations respond to the demands of an ever-changing environment. At any one instant we see the result of the interplay of these changes with such superimposed cycles as day and night, the rhythms of tides and phases of the moon, annual cycles, seasonal changes, and less-well-defined, longer-term cycles.

For a species to continue generation after generation, enough individuals of that species must successfully reproduce to maintain the population. Modern evolutionary theory has clearly demonstrated that the only measure of the success of an individual lies in its offspring: whether an organism is fit or unfit is revealed only by the reproductive success of its progeny, ten, fifty, or a thousand generations later, and even this is tempered by a very strong element of chance. Every sexually reproducing individual has its own unique genetic constitution that will never be duplicated. One of the fundamental consequences of sex is that hereditary materials are shuffled and recombined in each generation, so that there is virtually no chance that the same combination will appear again—ever. The hereditary combination that produced a failure this year might have been highly successful last year, or might be so next year, when the circumstances and environmental conditions were or will be ever so slightly different.

In our modern world of technology we are fond of thinking that things must be either "functional" or beautiful, and, all too often, utility is given precedence over aesthetics. How is it, then, that in the highly competitive world of nature there should be so much that man finds beautiful? Clearly any beauty in nature is not at the expense of function but a result of it, and there is no separation of aesthetics and practicality. The streamlined body and silvery-blue color of the mackerel superbly fit it for life in the open sea. The supreme harmony between the organism and its environment is the consequence of a system where anything short of perfection cannot long survive.

To understand the relationship between the organism and the environment, it is necessary to consider the organism first. Modern molecular biology has concerned itself with the details of the chemical and physical processes that all together are called life. A profound observation has resulted from this work: that all organisms—plant and animal, microbe and man—are fundamentally similar and differ more in complexity than in kind. This accounts for the basic finding that all organisms do the the same things, although they may do them in different ways.

In order for an individual to survive it must: have room in an environment where physical and chemical conditions are acceptable to it; find food, water, and oxygen, while avoiding predators and disease; be able to eliminate its waste products, such as carbon dioxide and nitrogenous compounds; and find mates (if necessary) and reproduce.

The sea is the home of all life, and those creatures that have left the sea to enter the terrestrial world have had to develop their own environment and insulate themselves against the hostile conditions, even as our space explorers must carry their insulation and environment with them. Desiccation, for example, is seldom a problem for marine organisms except those that live in the intertidal zone. Therefore they have no need for waterproof skins and impervious shells, except as defense against mechanical injury. There are, however, problems of water balance with which every organism must cope, and in one way or another every successful species has met this challenge.

All life processes require energy and, of course, building materials. Energy on this planet comes ultimately from the sun and is continually fed into our system. Some is lost by reflection and radiation; the rest is consumed or stored in chemical compounds. In contrast to energy, the minerals required by living organisms are in limited supply and they must be used over and over again. These important principles are readily illustrated in the sea. Plants, of course, are the only organisms capable of using light energy directly to build their own tissues from inorganic substances, and therefore, as the producers, they form the base of all food chains, being fed upon by herbivores, which are in turn eaten by carnivores, and so on. Ultimately all living things die and decompose, thus freeing essential minerals to enter the cycle again and again. The richest areas of the sea are those where critical materials are brought to the surface by upwelling currents or fed into the sea from the land. In such areas space, rather than food, may be the factor that limits the number of individuals present.

Every marine community has a definite structure as rigid as that of any human community. Within the community plants produce food materials from minerals and carbon dioxide and water and incorporate these materials in their own structure. The plants are then eaten by small animals, which in turn are eaten by larger and yet larger animals.

These food relationships alone would give some order to the community, but there are other social relationships that are equally important (one of these will be discussed in Chapter 8). Most communities consist of residents and wanderers. The residents spend their entire lives in one small area, never moving more than a few feet after they complete their planktonic larval stages. The wanderers, by contrast, come and go, spending a few hours or a few days at one place and then moving on to another location. Sometimes there will be definite migrations from one region to another for spawning or feeding, but most of the time the wanderers seem to drift along aimlessly wherever the needs of finding food and shelter take them. The more we study these complex social structures, the more we realize that man is not unique after all and that other animals have their social interactions too.

The oceans cover nearly seven-tenths of the earth's surface, and we now know that there is life even in the deepest parts, where sunlight never penetrates. Plants and animals that live in the ocean have developed a seemingly endless array of adaptations that enable them to take advantage of special features of the environment. Barnacles attach to solid rock; jellyfish float near the surface; herring and the whale shark feed on plankton. Many more of these adaptations are illustrated in our photographs or pointed out in the text. Each part of the ocean, such as the arctic, and the abyss, has its unique array of living creatures that coexist in a special kind of dynamic harmony, and the adaptations to any particular habitat must necessarily include mechanisms for avoiding direct competition with the other organisms that share that habitat. Even an artificial object such as a manmade piling provides an environment where starfish (sea stars), sponges, mussels, urchins, anemones, and algae soon form a special community. There is an inextricable interweaving of the physical, chemical, and biological factors of every environment.

The surface of the ocean may look pretty much the same everywhere; even when we put face masks on and look down we can't see very far into the water and we can't see subtle chemical differences that are very important to the organisms that live there. There are many factors that affect the distribution of marine organisms. Temperature, depth, currents, amount of wave action, turbidity, and chemical compounds are a few of the important features of the environment. Animals that live near the surface of the open sea must have a set of structural and behavioral adaptations different from that belonging to animals that live on the bottom or in shallow waters along the coastline. This difference is often manifested in colors. Open-water forms are mostly transparent or tinted with various shades of blue. Animals that live in the shallows tend to be brownish or sometimes brightly colored, and in deep water reds and black predominate.

The oceans must be thought of as consisting of a number of subdivisions. Biogeographers and ecologists have worked out rather elaborate classifications based on the distribution of plants and animals. Perhaps the most important factors that limit the distribution of living organisms are depth and temperature. In all oceans there are a near-shore and an open-water province, each with subdivisions which have characteristic assemblages of organisms at particular depths. It is also noteworthy that in the sea, as on land, temperate and cold-water environments have fewer species but more individuals than do tropic communities, which are characterized by a large number of species, no one of which is greatly more abundant than the others.

Thus, we find tremendous stands of kelp and large beds of mussels along temperate shorelines, and many different kinds of corals and fishes in tropical areas. Exactly what causes this difference is a matter of debate among ecologists; there is no obvious reason why the situation could not be the other way around. Books have been written on this subject, and a number of theories advanced, but as yet no definitive answer has been found.

We do know that no living species is found in all parts of the world; every organism has a definite native range. The presence of a particular species at a particular location is due first to its having had access to the area, and second to its having found suitable conditions for its continued survival after it arrived. The study of the patterns of distribution of living species is known as biogeography and it is one of the most challenging areas of natural history, involving intensive collecting in all parts of the world and a careful consideration of the organisms' ecological requirements in order to interpret the observed patterns. For many groups of organisms we have scant data on either distribution or ecology, and so biogeography continues to be an active field.

Land masses and temperature are the most obvious real barriers to marine organisms. We find distinct separations between the arctic, temperate, and tropical faunas, and we recognize that the species that live in the North Pacific are by and large quite different from those that live in the North Atlantic or the Antarctic regions, although many families and some genera occur in more than one of these areas.

In the tropic zone there is a clear distinction between the vast Indo–Pacific fauna, which ranges from the east coast of Africa eastward to the eastern limits of Polynesia, and the Atlantic fauna. Interestingly enough, the fauna of the west coasts of the Americas, although distinct, shows a closer affinity to that of the Caribbean–Atlantic region than to the Indo–Pacific fauna. This suggests that the open sea has been more of a barrier than the Central American land bridge, which has been elevated and submerged several times in the geologic past, making it possible for sea animals to cross from one ocean to the other without difficulty. The vast stretches of open ocean in the eastern part of the Pacific basin offer no shallow-water stepping stones to permit those species that live in shallows to cross these great distances.

Many of the animals shown in this book live on coral reefs. There are several different kinds of coral reefs in the world, but they are all restricted to the tropics and subtropics between 35 degrees north and 32 degrees south latitude. Coral reefs are generally conceded to be a diver's paradise because corals thrive in warm, clear waters, and there are so many kinds of organisms around a reef that most of us never spend a half-hour in the water without seeing something that we have never seen before. Coral-reef structures have been built by living organisms that extract calcium carbonate from the sea water and incorporate it into their skeletons. While corals are the most conspicuous and spectacular of the reef-builders, other organisms, such as sea urchins, mollusks, even single-celled protozoa, and especially certain plants also contribute to the reef framework. At the same time that these organisms are building the reef, still others are busy destroying the skeletons of the reef-builders, and the end result is that the reef community acts like a kind of superorganism with an over-all organization and metabolism that can be compared with that of a single individual. To carry our analogy further, it will be seen that a reef can even reproduce itself, for if a new volcanic island were to arise nearby it would sooner or later be colonized by larval stages drifting over from the old reef, and eventually a new reef would be formed with a structure similar to that of the "parent" reef.

The analogy between the organism community and a single individual is an oversimplification, but it teaches us a number of important natural principles. First of all, the community, like the individual, must have a proper environment. Coral reefs can grow only on firm rock bases in warm waters in areas that are relatively free from silt and away from fresh-water runoff from the land. The rocky shores of New England can support one association of animals; the coastal marshes of Georgia or Louisiana can support another. Next, the community must have a supply of minerals and energy. Energy from the sun recycles within the community, and the minerals are also used over and over again. Finally, the community is a self-regulating system that maintains a delicate balance among its various members. There are many mechanisms that operate simultaneously to regulate the populations in the community, just as there are numerous mechanisms for maintaining water balance in the individual. At the present time vast expanses of reef in the western Pacific Ocean are being destroyed by the crown-of-thorns starfish, which for some reason has undergone a dramatic population explosion in the last few years. Apparently the natural checks on this predator have somehow been upset and it has become so abundant that it is literally eating itself out of house and home. We do not know why this has happened; some scientists have speculated that long-lived pesticides may have killed off a natural predator of the starfish and allowed it to overpopulate. Whatever cause is eventually discovered, the crown-of-thorns will always be an outstanding example of what can happen when the natural regulatory processes in the community are disrupted. It also demonstrates how little we know of the mechanisms that regulate animal populations.

Men are fascinated by the sea and recognize it as both friend and enemy. Today, as modern technology allows us to explore beneath the surface even to the deepest parts of the sea, the challenge it provides is greater than ever. It is somehow comforting to realize that the most sophisticated deep-submersible vehicle holds the same promise for the exploration of the ocean floor that the first crude sailing canoes did for the surface, and that even now there are frontiers to be met and challenged on this battered old planet.

The sea has always been more than just water. It is a source of food, a route for travel, a feared and destructive enemy, and a wide sewer into which to dump our refuse. It has also been a source of beauty and comfort that has brought tranquillity to men's souls. Today, more than ever before, we turn to the sea for relaxation. We sail our yachts on it, we wade and swim in it, we explore it with our aqualungs, and, perhaps most important of all, we watch its ever-changing moods and we feel better knowing it is there—scarred here and there from its battle with mankind, but basically triumphant and unconquered. Instinctively all men see in the ocean hope for the future.

As the land becomes more and more crowded, men look to the ocean to provide their needs; food, minerals, even living space must come from the sea. Our exploitation of the sea is still very much in the exploratory stage; we don't have enough basic information to be good hunters, and it is no more reasonable to expect to feed the world population from hunting in the sea than it would be to expect to feed everyone by hunting on land. Eventually we must learn to farm the sea as we have learned to farm the land.

Full exploitation of the sea can be realized only on the basis of a thorough knowledge of marine-organism communities. We must learn to manipulate populations so as to encourage the most valuable and useful species, and we must learn to harvest these species in ways that will permit the largest catch that is consistent with maintaining the population for the future.

At the same time we cannot neglect the important aspects of conservation. Some areas must be set aside to be left untouched for future reference and basic research. Other areas must be maintained for recreational as well as commercial use. Prominent among the preferred sites for recreational diving and sightseeing will be the coral-reef areas of shallow tropical and subtropical seas. However, every marine environment is important as an element of the total. It is imperative that man understand his relationship with his environment and live without destroying it. Man sometimes forgets that he is as involved with his environment and as subject to its laws as other organisms. He cannot destroy his habitat without destroying himself.

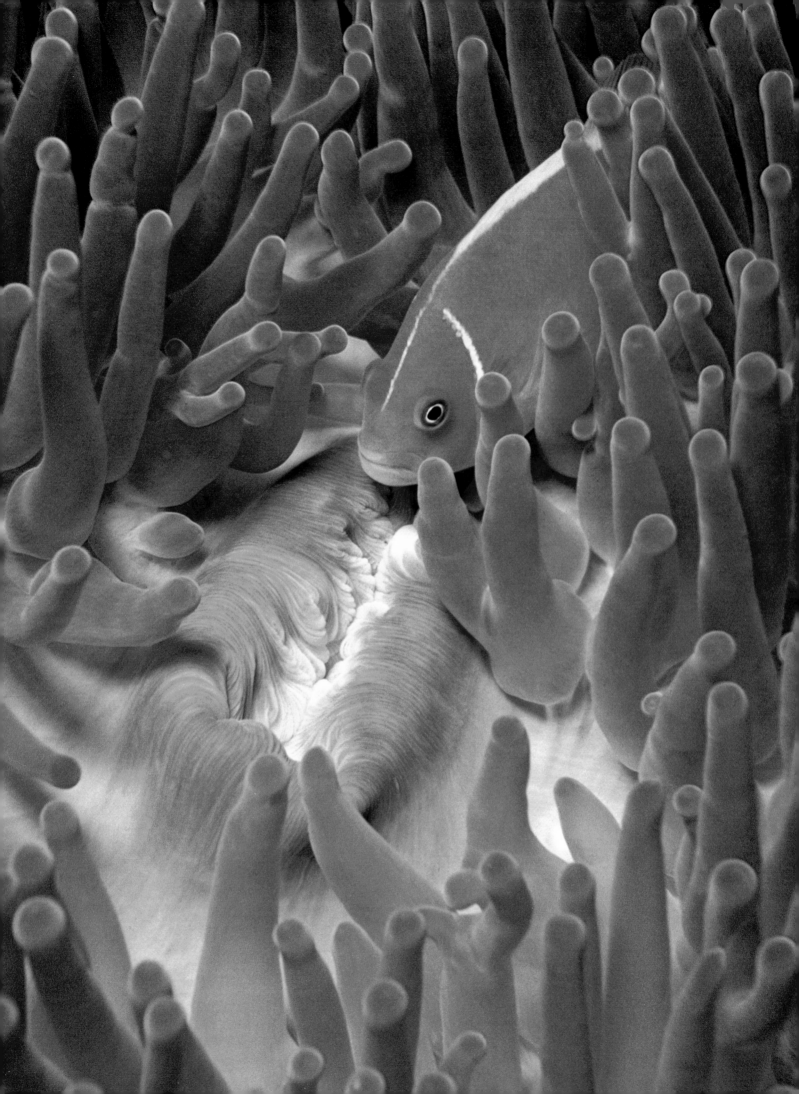

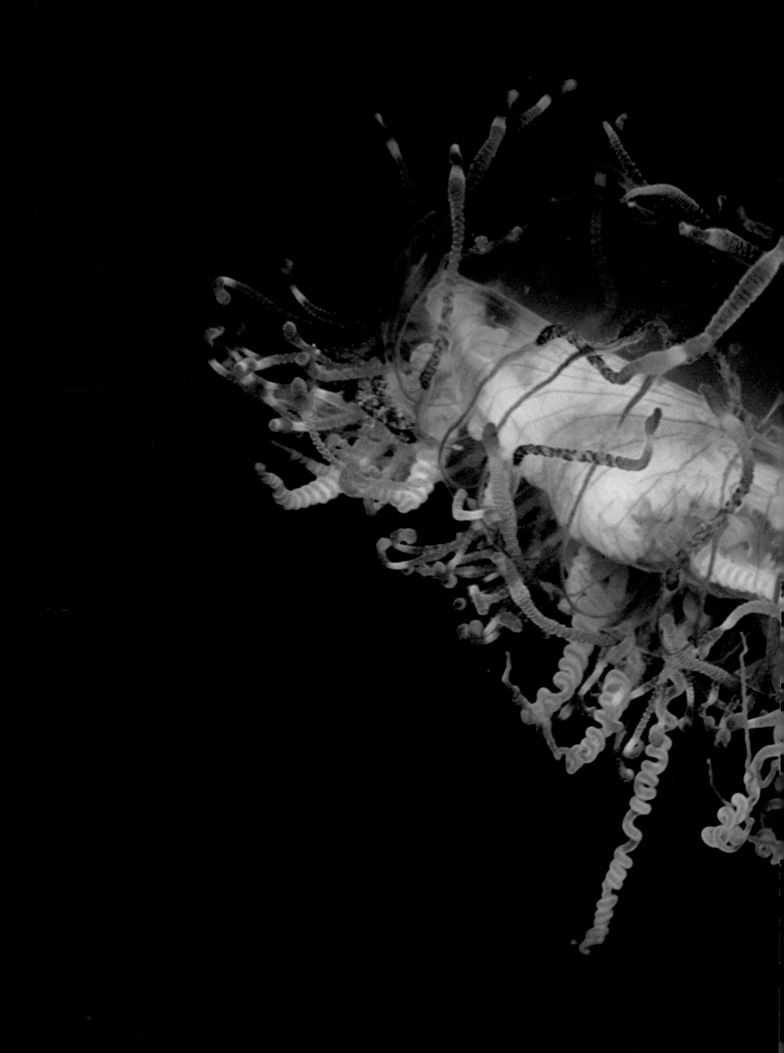

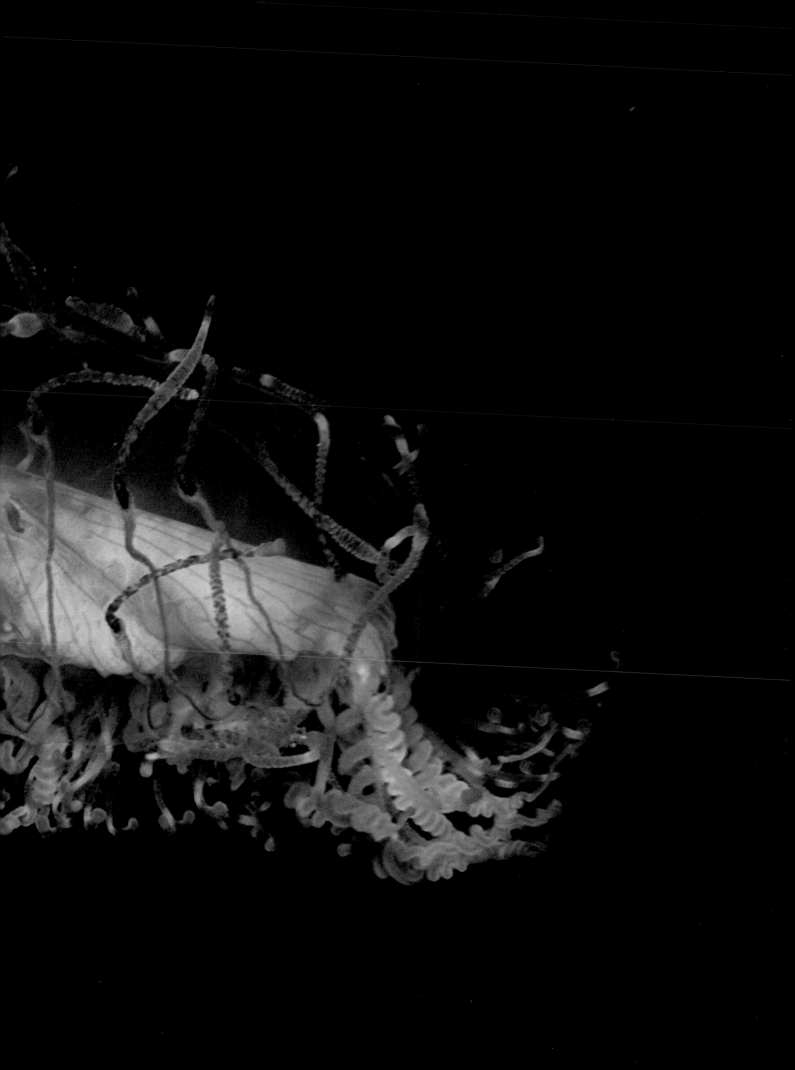

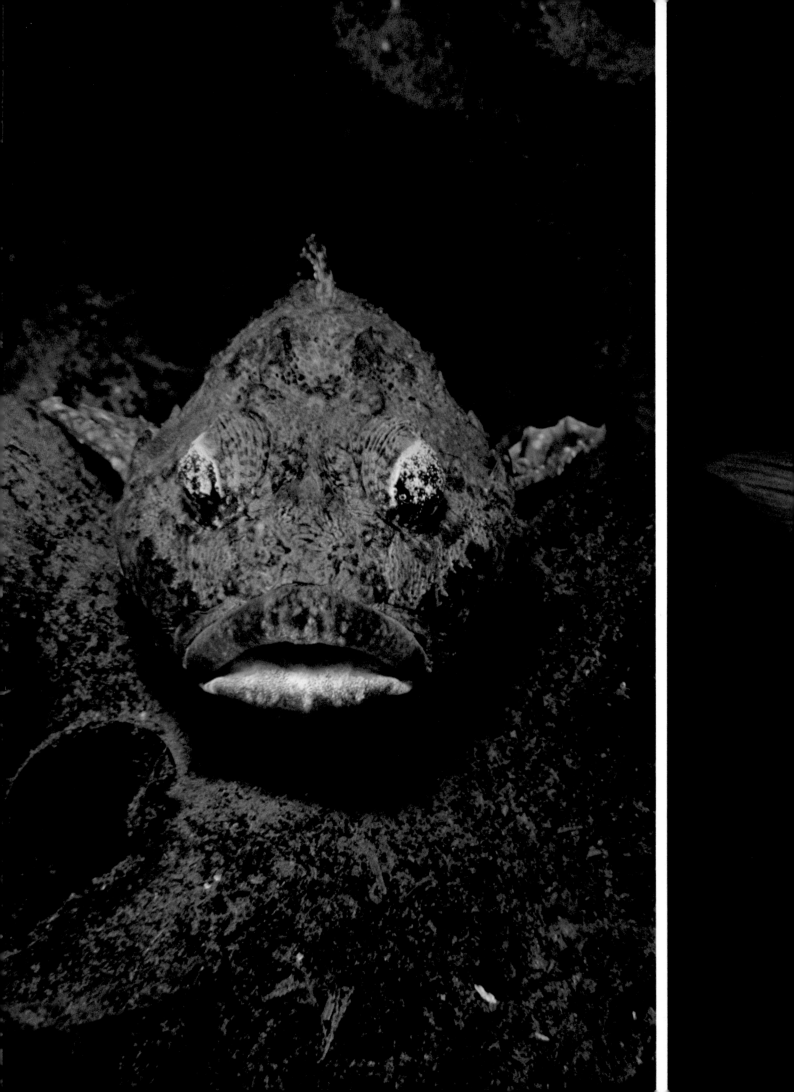

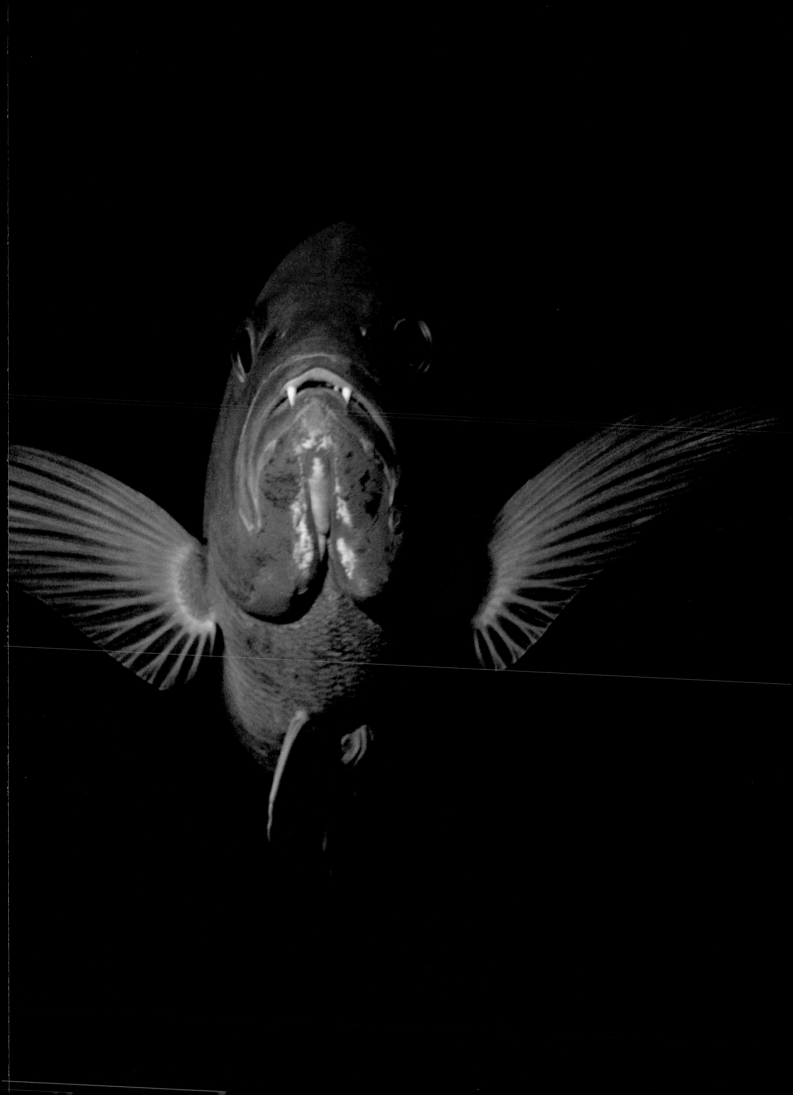

2. SPONGES

Nearly everyone is familiar with the few kinds of sponges that are used for washing automobiles and people, but the skeletons we know in the bathtub bear little resemblance to the living animals and give no hint at all of the marvelous diversity of the sponge phylum Porifera. Natural sponges have become something of a rarity now that cheaper and "almost as good" synthetic products have become readily available. The fact that there are only a few inconspicuous fresh-water sponge species has kept most inlanders from appreciating this varied and fascinating group of living organisms.

Sponges are animals, but they have a curious, rather informal organization, somewhere between the one-celled protozoa and the more rigidly structured higher animals. Thus they present something of a paradox. In some ways they act as a group of single cells organized into a loose alliance. In other ways they are highly organized, like the multicellular animals. Recently it has been discovered that some sponges, at least, lack RNA, one of the basic materials of the genetic system of most plants and animals.

Unlike the cells of higher animals, the cells of sponges are not organized into tissues and organs such as stomach, liver, brain, and so on. Instead, there are several cell types, arranged in two layers surrounding a central cavity and separated from each other by a gelatinous material called mesenchyme. The outer layer consists of hexagonal cells, and the inner layer is made up of cells with an elongated whiplike flagellum surrounded by a collar-like rim. These collar cells are very much like certain protozoans, and it is believed that sponges arose from such protozoa.

The outer wall of the sponge is penetrated by numerous small openings, each surrounded by a tubular cell called a porocyte. Water moved by the corkscrew action of the flagella enters the sponge through these tiny openings and is then swept through the sponge and out of a larger opening called the osculum. The flagellated cells also trap food particles and either digest them or pass them on to other amoeba-like cells lying within the gelatinous middle layer. Sponges have no nervous system.

The skeletons of sponges may consist of either small rods of calcium carbonate or silica, called spicules, or fibers of a protein called spongin. Some sponges have both spongin fibers and calcareous spicules, but commercial sponges have only a spongin skeleton, and this, of course, is what we see in the finished product. The spicules, which are microscopic in size, come in a variety of sizes and shapes, and in order to identify some species of sponges it is necessary to boil a bit of the tissue in acid to free the spicules so that they can be examined under a high-power microscope. Some spicules are simple needles, some are star-shaped, and some look like miniature grappling irons or jackstraws. Robert R. Newport

Sponges reproduce either sexually or asexually. There are no special reproductive organs (or any other organs, for that matter) and the eggs and sperms develop from amoeba-like cells in the mesenchyme, or from collar cells. Some sponges are hermaphroditic, each individual producing both eggs and sperms; others are distinct male and female individuals. The sperms are released into the water, are carried by the currents to other sponges, and eventually reach the eggs. Early sponge development takes place in the mesenchyme layer of the parent, and at the appropriate time a free-swimming embryo breaks out, swims about for a while, then attaches to the ocean bottom and proceeds to develop into a new individual. Different groups of sponges have somewhat different larvae, but the general process is essentially the same. When the larvae are still in the parent sponge they can be seen when one cuts through the sponge. Young larvae are whitish, about pinhead size, and older larvae are greenish-black.

Because sponges are so loosely organized they have great powers of regeneration. Some sponges can be squeezed through a cloth filter so that the cells are separated, and then, if they are left undisturbed, the cells will reassemble and continue to grow. This power of regeneration makes it possible to cultivate commercial species by cutting large sponges into small pieces and tying these pieces to rocks, then sinking them in places where conditions are favorable and allowing them to grow into complete sponges. The quality of the sponges varies with the conditions under which they grow and can therefore be controlled by planting the sponges in the proper environment. Furthermore, the sites can be selected so that the sponges will be easy to harvest. Very little is known about the biology of the sponges except for those species that are commercially valuable and a few other relatively simple ones that are commonly used in classrooms and laboratories.

The basic form of sponges is a simple closed tube, but this arrangement limits the size of the sponge because the weak pumping power of the flagella of the collar cells is not adequate to move a large volume of water through the central cavity. The efficiency of the system is increased when the inner layer is folded so that there are many more collar cells in the same volume of sponge. This folding also reduces the volume of the central cavity and permits a more efficient flow. Many sponges have the outer walls folded as well. Instead of one large osculum, they have many smaller excurrent openings all over the body, so that the original symmetry of the tubelike design is obscured completely. The tubular or vaselike form of some of the sponges illustrated in this book is a secondary development, and some of the larger openings that show in the photographs are not equivalent to the excurrent oscula of simpler sponges.

Sponges must remain attached to a proper firm bottom except during the short period when they are free-swimming larvae. It is thought that most commercial species spend only a few days as larvae, but the presence of sponges around such remote islands as Bermuda and Hawaii suggests that some must have longer larval periods, during which they are able to drift across hundreds of miles of open ocean to a suitable habitat. Once they become attached, they cannot pack up and leave again. Those that do not settle in a favorable situation are doomed to perish. Most marine sponges cannot tolerate low salinities. Sponges do occur in temperate and arctic waters, but they are most conspicuous in the tropics.

The classification of sponges is based mainly on their skeletal structure. The three major groups (classes) are the Calcarea, which have calcium carbonate spicules, the Hexactinellida, which have six-pointed silica spicules, and the Demospongiae, which have either two- or four-pointed spicules or spongin fibers or both spicules and fibers. The Demospongiae include the commercial sponges and most of the species that live around coral reefs.

In addition to the cushion-shaped and vase-shaped sponges, there are some that have a branching form. Others are encrusting species that form a thin layer like a soft blanket on hard rock bottom.

Sponges come in a variety of colors. A few, such as the stinker, are a drab grayish brown, and some of the commercial sponges are jet black. Others, as seen here, may be brilliant yellow, orange, or crimson. Some of the tubular deep-water species fluoresce and glow in the dim light at 100 or 150 feet below the surface.

In most coral-reef areas sponges play an important role in determining the ultimate structure of the reef, not only by adding their substance to it, but also through their attacks on the skeletons of corals and other reef-dwelling organisms. Boring sponges are able to drill through shells and even solid limestone. Exactly how this is done is not known, but it has been shown that boring sponges are extremely abundant on the slopes beyond the front of the active reefs, where they attack coral colonies that have been broken off and have tumbled down the slope. In many cases they may be responsible for weakening the bases of the colonies in the first place. Although sponges are sessile and cannot run away from their enemies, they are far from defenseless. They have sharp-pointed spicules that discourage many predators, and some, such as the fire sponge, *Tedania ignis,* produce a chemical that causes considerable skin irritation to people who touch it, a dermatitis that has been likened to the effects of poison ivy. The stinker sponge, a common species in shallow waters off the Bahamas, has an extremely disagreeable odor. Nevertheless sponges are fed upon by some fishes and make up over 85 per cent of the diet of the West Indian French and Gray angelfishes.

Sponges, like other animals, are affected by diseases from time to time. Usually epidemics go unnoticed, but in 1938 and 1939 commercial sponges in the Gulf of Mexico and throughout the Bahamas suffered outbreaks of a disease believed to have been caused by a fungus, although this is not at all certain and other factors, perhaps other organisms, may have been involved. Between 90 and 95 per cent of the commercial sponges were destroyed, and the velvet sponge, a desired species, disappeared from the Bahamas and has not been reported from there since. Another outbreak of disease occurred along the west coast of Florida in 1947 and 1948, but again the culprit was not discovered. West Indian sponge populations are now building up, but the industry has never recovered and synthetic sponges have supplanted the natural product for most purposes.

Some sponges have been described as veritable apartment houses, and one of the favorite pastimes of field marine biologists is examining sponges to see what animals have set up housekeeping inside the chambers and canals of the sponge. Pistol shrimps are often found in West Indian sponges in unbelievable numbers. One ecologist counted more than sixteen thousand in one large sponge. Worms, sometimes several feet long, wind themselves around and around inside some sponges, and crabs burrow into them, much to the disgust of commercial sponge fishermen, for the crabs occupy large holes which greatly reduce the value of the sponge. Apparently the crab does not actually eat the sponge, because the holes are smooth and covered with living flesh of the sponge, but the presence of the crab prevents the sponge from filling in the hole, and the sponge probably grows around the crab. Some algae grow on certain sponges, and many clams, snails, limpets, and brittle stars also live in and on sponges. But the most remarkable residents are fish, some of which are small and slender and are not known to live anywhere except in sponges. They even lay and guard their eggs within the recesses of the sponge. William Beebe told of collecting small fish from tube sponges in the West Indies by plugging the openings under water and then sending the sponge to the surface, where the trapped fish were removed and soon became scientific specimens.

One of the silica sponges which has its spicules fused to form a lattice-like structure well deserves its common name, Venus's flower basket. Certain shrimps live in this as commensals: a male and female enter when they are small and become trapped as they grow too large to escape. In Japan the dried sponge and its inhabitants were often presented as wedding gifts symbolizing lasting "togetherness."

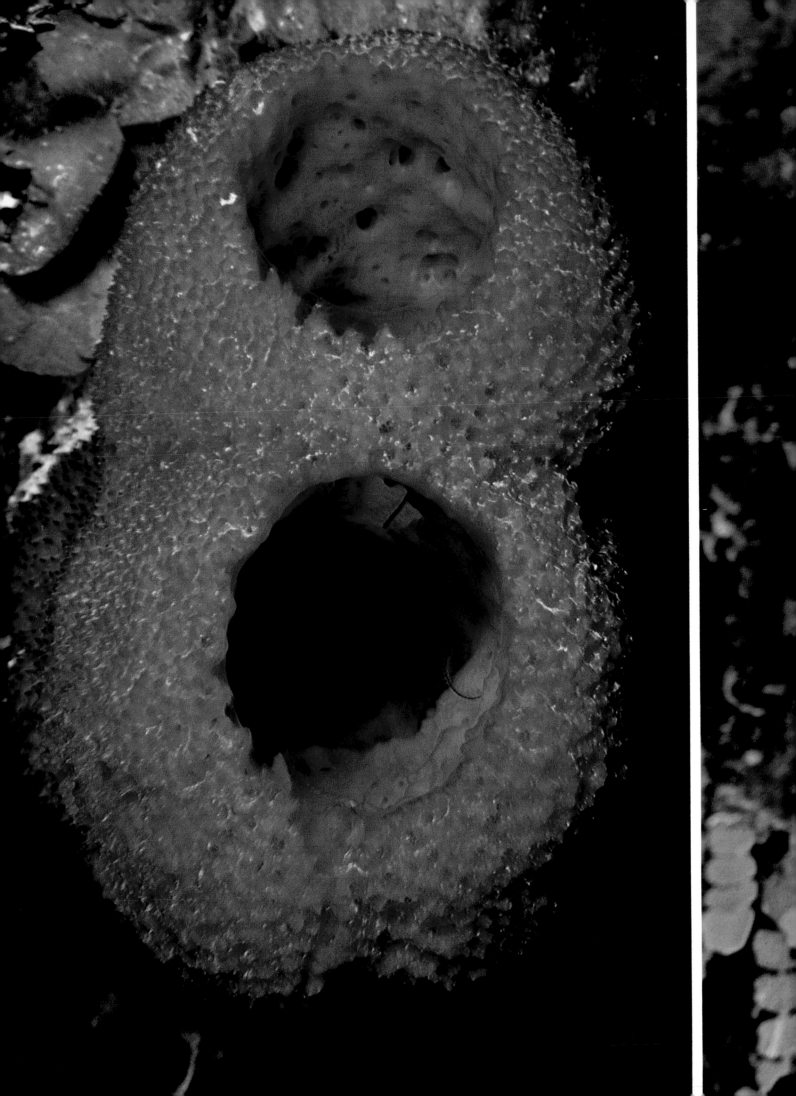

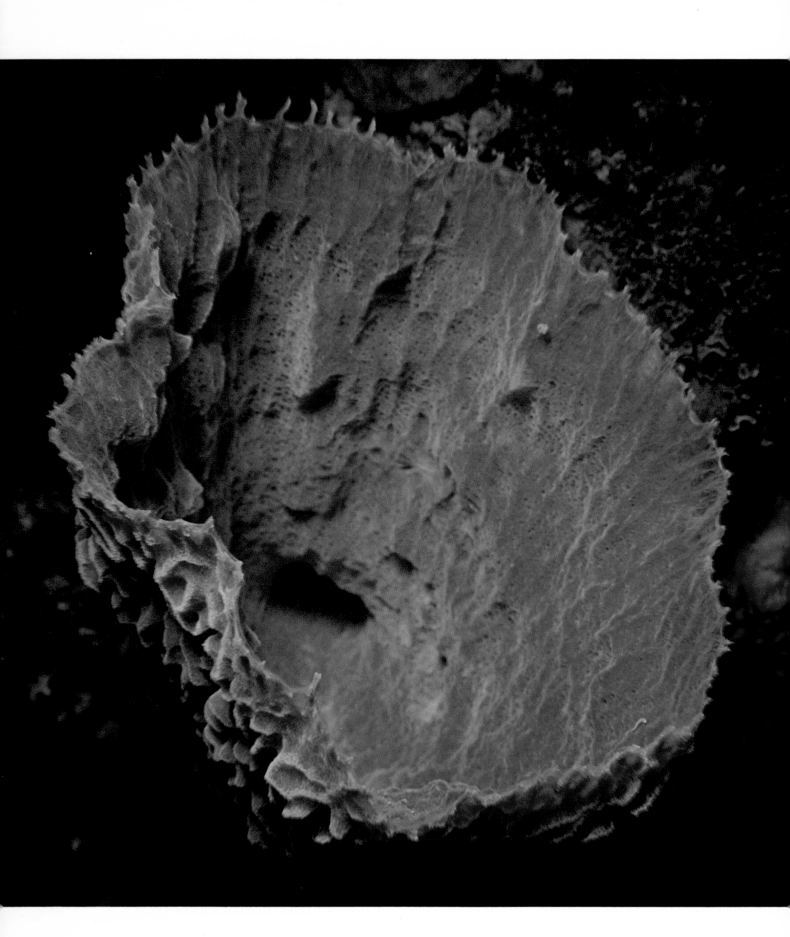

3. CORALS

Corals are among the most fascinating organisms in the sea. No collection of marine souvenirs is ever complete without at least one delicately sculptured skeleton of a reef-building coral. Few objects can so readily call forth dreams of adventure in far-off places. But, lovely as they are, the dry white skeletons of the souvenir shops are a far cry from the living organisms in the sea, and they can never fully tell the story of this fascinating group of animals.

Corals are animals, but they are many times mistaken for plants since they so often look like bushes covered with tiny, delicately colored flowers. The objects we find dried and bleached are not, of course, the whole animals but merely the limy skeletons that the animals secrete around their bodies. The living tissue, which is often brightly colored, is confined to a thin outer layer that scarcely conceals the form of the skeleton. Most coral skeletons are constructed by colonies of hundreds of individuals working together to produce a highly organized and complex structure.

Coral colonies start when one or more larvae settle on firm bottom, on a rock outcrop, or on another coral colony, become attached, and begin to grow. They assume many shapes: some are treelike, some are hemispherical masses, and some merely encrust the bottom. Each species has a set pattern, but this can be modified somewhat by conditions at the time it is growing.

Under some circumstances a globose coral colony can increase its diameter at the rate of about one-tenth of an inch each year. This, however, must be near maximum. As the colonies grow, they provide a base for the growth of other colonies and in this way build a coral reef. Wave action and coral-eating animals such as sponges, mollusks, and parrot fish break down some of the colonies and grind their skeletons to fine powder that settles into crevices, where bacteria, plants, and chemical processes cement the coral sand together. Slowly the reef becomes an almost solid structure, like a mass of concrete.

Upward growth of corals is limited by the surface of the water, because coral cannot withstand prolonged drying. This is why coral reefs are so hazardous to navigation; they lie just under the surface except during low spring tides.

Corals belong to the phylum Coelenterata, which contains three classes: the Hydrozoa, the Scyphozoa, and the Anthozoa. The Hydrozoa class encompasses the familiar fresh-water hydra as well as the so-called fire coral and the highly complex Portuguese man-of-war. (The latter is actually a colony of many individuals, some of which are specialized for feeding, some for reproduction, still others to form the beautiful blue gas-filled float and sail that allow the colony to cruise at the surface of the ocean.) The class Scyphozoa includes the jellyfish, which get their name from the massive gelatinous tissues that form the bulk of the body. The Anthozoa are the soft and the stony corals and their close relatives, the sea anemones.

Coelenterates are a little more complicated than sponges. Their body walls consist of an outer and an inner layer, each composed of several different kinds of cells. Between these layers there is a middle layer called the mesoglea that may be either a thin membrane or a thick and jelly-like mass. Some of the cells act as muscles and by contracting and relaxing cause the body and especially the tentacles to move. Other cells perform special functions such as secreting mucus, producing enzymes, and acting as sensory receptors and nerve cells. Of particular importance is a characteristic type of cell that contains nematocysts, the weapons of the coelenterates. Some nematocysts secrete a sticky material which is used to attach the polyp to the bottom. Others produce threads that entangle the prey, and still others are tiny tubes that penetrate and inject a poison into the prey or for defense against an attacker. In the undischarged state, the coral's nematocysts are contained within the surface cells, ready for action. A trigger-like point that can be activated by physical contact, or sometimes chemically, causes the nematocysts to shoot out. These minute cells are tiny but potent. They are responsible for the stinging caused by the Portuguese man-of-war, the fire coral, and the dreaded sea wasp. All coelenterates have nematocysts, but only a few are really powerful enough to irritate or affect a human being seriously.

The basic coelenterate structure can be seen in the individual coral animal, called a polyp. Each polyp is essentially a tube, closed at the lower end and with a mouth opening at the top. The internal cavity, where digestion takes place, is called the gastrocoel ("stomach"). The mouth is surrounded by tentacles—four, six, eight, or multiples of these.

The individual polyps of stony corals are generally much smaller than anemones. Although some, such as the mushroom coral, may be as much as 10 inches across, most are on the order of 1 to 3 millimeters, generally less than one-eighth of an inch. The mushroom corals consist of a single large polyp, but most colonies contain thousands of tiny individuals. The polyps of some sea fans are so small as to be scarcely visible to the naked eye. These colonies are more than mere aggregations, for each individual is physically connected with adjacent polyps, and the general design of the coral colony is, within certain limits, determined by the hereditary make-up of the particular species of coral.

Coral grows by incorporating calcium carbonate out of the surrounding water into its skeleton, which is laid down around the base of the polyp. The base of the polyp is usually folded, and this produces a series of radiating walls within a cuplike depression. The number of these walls and their size and shape are used as identifying characteristics of coral species. In the very large polyps of the mushroom coral (Fungia) the stony walls look like the "gills" of a mushroom, and the scientific name reflects this resemblance. The so-called brain corals have their polyps arranged in rows with adjacent polyps not separated by skeletal material. The rows, however, are separated from one another, and this gives the structure a convoluted appearance not unlike that of the outside of a human brain.

Soft corals, including the sea fans and sea whips, are arranged somewhat differently. Each colony also has a horny central rod that supports it and gives it its shape but is flexible enough to permit the colony to bend with the currents. Surrounding this central skeleton is a thick matrix in which the polyps are embedded. The matrix is penetrated by interconnected tubes which are extensions of the central cavities of the polyps. The matrix also contains calcareous spicules. The precious black coral that is used in making jewelry is cut from the axial skeleton of a species of soft coral.

Sea fans, another kind of soft coral, grow in a most interesting fashion, branching and then reuniting to form a latticework with regular perforations, usually in a single plane so that the colony is quite flat.

Coelenterates reproduce both sexually and asexually. In the latter process, called budding, a section of the body wall grows outwards as a pouch that eventually develops tentacles and a mouth on its outer end. Finally the bud separates as a new individual or, in colonial forms such as the corals, remains attached but functions independently.

Sometimes sexually reproducing generations alternate with asexual generations. When this is the case, the generations may differ radically in appearance, the asexual generation being polyp-like while the sexual generation is a free-swimming umbrella-shaped medusa. The familiar jellyfish are large medusae. In some groups of coelenterates one generation or the other has been eliminated. The corals, for example, have no medusa stage, and the polyps produce the eggs and sperms.

Anemones have a wide geographic range and are conspicuous in cold waters as well as in the tropics. Although anemones are sessile animals, they can move about by gliding along on their bases, and some reproduce asexually by a fascinating method called pedal laceration. As the anemone glides along, pieces of its base are left behind and grow into new anemones. Some have very long delicate tentacles, and an aggregation of anemones can easily be mistaken for a group of plants. A number of fishes, crabs, and shrimps have evolved an intimate relationship with certain anemones and can live among their tentacles unharmed, while other animals are stunned by the nematocysts of the tentacles and are eaten by the anemones.

Coral polyps feed on small animals, mostly planktonic forms that drift with the tides and currents. The extended tentacles seize their prey after it has been stunned by the nematocysts, and pull it into the central mouth. Inside the stomach, enzymes break the food into small particles that can be engulfed by the surrounding body cells, where the final phase of digestion takes place. Later, indigestible materials are merely expelled through the mouth.

Through the processes of evolution the feeding cycles of the corals have become adjusted to critical factors in the environment. Some species feed at night, when the plankton is concentrated near the surface. Other species feed when the tides and currents sweep food to them. Because the plankton organisms tend to sink and scatter in deeper waters during the day, deep-water corals are likely to feed actively at that time.

Although stony corals also live in cold waters and in deep parts of the ocean, their greatest development is in the tropics, and the reef-building corals are, for all practical purposes, restricted to the tropical and subtropical zones.

Reefs develop best around low islands because most corals are unable to tolerate either silt or fresh water, both of which are present in large amounts near the continents. Furthermore, reefs are unable to flourish in water deeper than about 300 feet, although individual coral colonies occur at greater depths. We now know that all reef-building corals have algae, single-celled plants, within their tissues. The corals get some of their nourishment from wastes given off by the algae, and the algae which must have light to grow actively, use the carbon dioxide and perhaps other waste materials from the coral for photosynthesis.

It appears that the only difference between the corals that build reefs and those that do not is the presence of algae in the reef-builders. Some very exciting experimental work is being done on this relationship. The algae can live outside the corals—in fact, part of their life cycle normally takes place out of the coral—and the corals can survive without the algae. When the algae are present, however, the rate of growth is about ten times as fast as it is when the algae are absent.

Species of coral differ in their requirements, with the result that they congregate in different areas, and this determines the basic organization of the reefs. In West Indian reefs the treelike elkhorn coral, *Acropora palmata*, thrives best where there is the most wave action. The largest colonies are at the reef front, where there is almost constant surf. The more delicately branched staghorn, *Acropora cervicornis*, is found in quieter and deeper waters beyond the front of the reef, while the back reef will be dominated by the curly, flattened colonies of lettuce coral. Most corals are sensitive to sediments; they secrete copious slime that traps the sediment particles and then is swept away by tiny cilia (hairlike processes) of the surface cells. Nevertheless, if the local sediment becomes too dense, it will result in the death of the coral, because most corals cannot move to a new location. A few coral species, however, are able to survive in muddy places. One of these, the coral *Manicina areolata*, has a special mechanism for survival. Like all other corals, it is attached to the bottom when it starts to grow, but later it breaks free and can move about, albeit very slowly. It can uncover itself if it is buried, and it can turn itself upright if it gets turned over. Mushroom corals in the Indo–Pacific can also right themselves. One species of solitary coral has solved the transportation problem in a unique manner. A sipunculid worm resides in a hole in the coral skeleton, and when the worm wants to move it has no choice but to take the coral along. The worm keeps its home, and the coral gets free transportation.

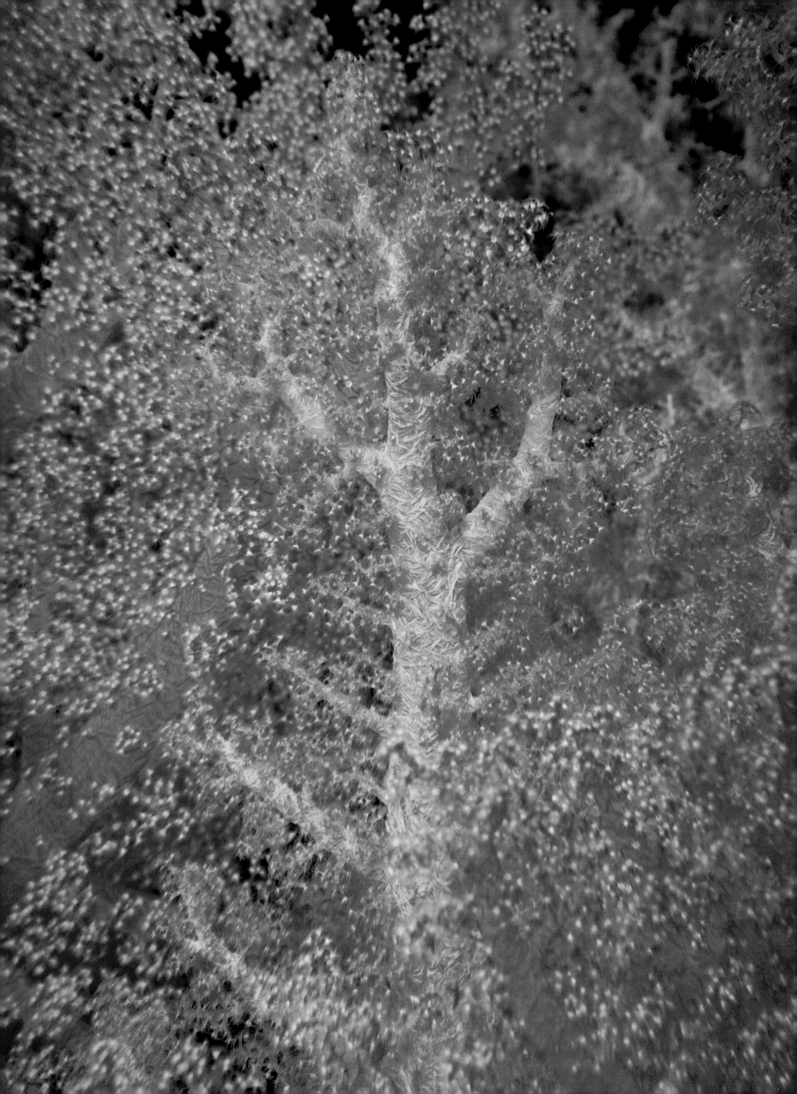

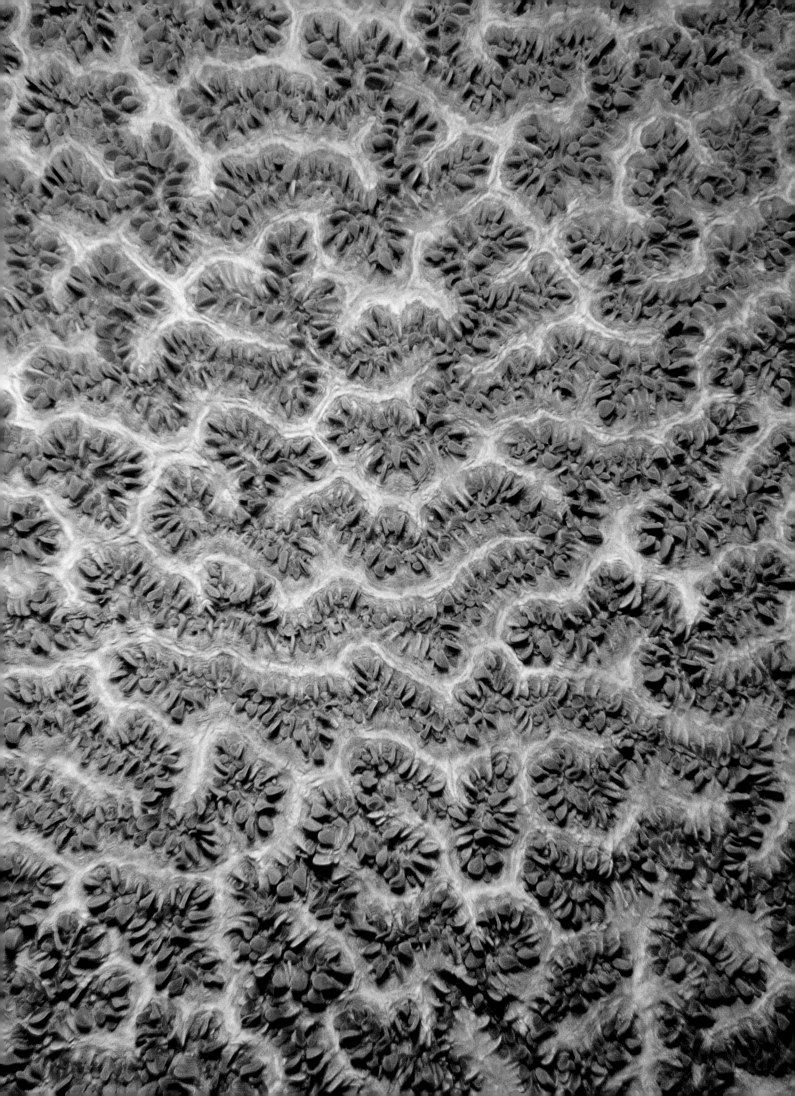

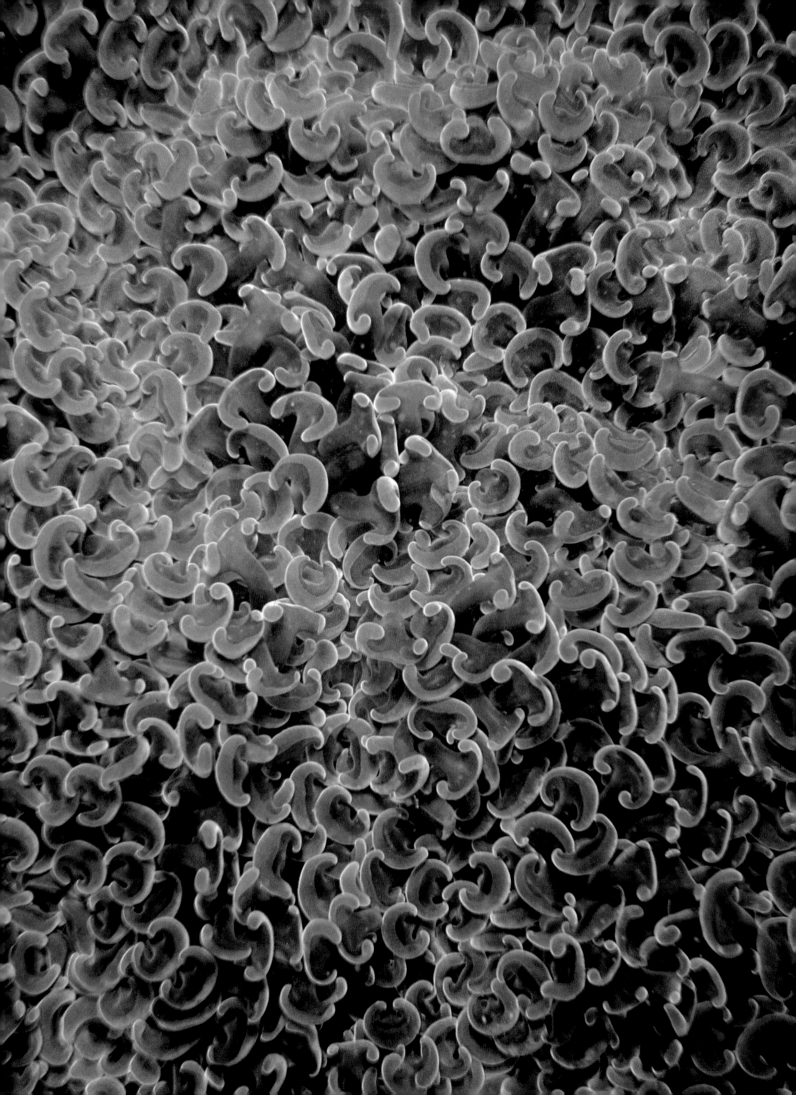

4. MOLLUSKS

Perhaps no natural objects are more reminiscent of the ocean than sea-shells. Certainly every nautical motif contains one or more shells, and few families ever visit a beach without returning with an assortment of these treasures. No seafood dinner is worthy of the name unless it includes at least clams, oysters, or scallops, all of which are representatives of the phylum Mollusca.

This is the second largest phylum in the animal kingdom, with more than eighty thousand species, and the life histories of its members are correspondingly diverse and interesting. Moreover, because of their hard shells, mollusks have been well preserved in the fossil record and a great deal is known of their history in the geologic past.

There are five major classes of mollusks that the diver is apt to encounter: Amphineura, the chitons; Gastropoda, the snails; Pelecypoda, the clams; Scaphopoda, tusk shells; and Cephalopoda, which includes the octopus, squids, the chambered nautilus, and ram's-horn shells. There are a few other classes of deep-sea or hidden forms that are of great interest to specialists but would hardly be recognized as mollusks by the casual observer.

Unlike sponges and corals, mollusks are bilaterally symmetrical, with a left and a right side and a more or less well-defined head. They have an upper or dorsal side, which usually has one or more shells, and a lower or ventral side, which is often thickened into a foot that can be used for moving about. The shell is produced by a thick membrane called the mantle. Usually the mantle overhangs the body so as to produce a space, the mantle cavity, which houses the gills. Thus the delicate gills are protected by the mantle and the shell. The digestive system is a tube, a drastic improvement over the blind sacs of the coelenterates, and most mollusks have a rasplike radula—a sort of tongue with many rows of teeth that can be extended for scraping food from hard surfaces. There is a primitive heart, which pumps clear blood through a few large vessels into some open spaces called blood sinuses. (A few mollusks have hemoglobin, and their blood is red in color.) The mollusk nervous system is primitive, with a few pairs of ganglia and varying degrees of sophistication of circuitry, from a relatively simple scheme in the chitons to the highest development in the fast-moving Cephalopoda. Cephalopoda such as the octopus and the cuttlefish have well-developed eyes which are strikingly like human eyes in appearance and function, although they are organized along fundamentally different lines. This coincidence has been hailed as one of the most notable examples of convergent evolution in the animal kingdom.

Because the mollusks are such a diverse group, there is hardly any environment that does not have at least a few molluscan species living in it. We are all familiar with land snails, and most fresh waters abound with snails and clams, but it is in the ocean that we find the most molluscan types. The skin diver will pass over the rather plain-looking tide-pool snails on his way to the reef, but he may be fascinated by the chitons that clamp tightly to the rocks at, or just above, the high-tide line. Chitons are one of the most primitive molluscan groups, with a shell that is divided into eight plates. Most chitons are small and dull in color, although one species grows to be a foot in length. They are not exactly speed demons. Scientists watched one individual for nine months, and it did not leave an area six feet square. Sometimes it did not move for weeks at a time.

Much more interesting are the Gastropoda, a group that includes the seemingly endless array of snails and derivative forms such as sea hares, limpets, sea butterflies, and most beautiful of all, the nudibranchs. Snails range from very tiny ones to the large conchs and whelks. Most are herbivorous, rasping plant materials off with their radulas, but some are carnivorous or even parasitic. The coneshells and the tulips are well-known carnivores. The coneshell preys on small fishes, harpooning them with a modified radula tooth that injects a virulent poison into the victim, which is then pulled back into the mouth of the coneshell by the radula. Many carnivorous snails feed on other mollusks such as clams and actually drill through the shells of their victims to inject a poison that relaxes the victim's muscles, allowing the shells to open.

Not all snails fit within their shells; in fact, some are so much larger that they completely surround the shell and get no protection at all from it. Then there are the cowries, which have highly polished shells that are favorites of many collectors. In these, the mantle extends out over the shell. The beautiful little flamingo-tongue belongs to this group, and it owes its striking pattern of orange spots surrounded by black rings to pigment patterns in the external mantle. The shell itself is plain cream-color. Other cowries have less colorful mantles, but their shells are exquisitely colored. Flamingo-tongues are predators of sea fans, rasping off the outer materials with their radulas, leaving the skeleton behind them exposed.

Snail eggs come in a variety of forms: some in gelatinous strings that look like a tangled skein of yarn, some in masses, and some in flat horny capsules. The moonshells lay their eggs in flat bands of sand grains cemented together and coiled into rings called sand collars.

Many snails are nocturnal and spend the day burrowed in the sand; others hide under dead coral rocks. Experienced collectors prowl the shallows at night, when the animals are active, or turn over rocks during the day to catch the snails in their hiding places.

Some gastropods have separate sexes, although many others are hermaphroditic with a single gonad that produces both eggs and sperms. Usually, the hermaphroditic forms copulate anyway and exchange sperms, so that there is cross-fertilization.

The snail's first line of defense is, of course, the shell, into which it can withdraw whenever it is threatened. Many snails have a door, called the operculum, that closes the entrance of the shell when the animal withdraws into it. In some this is a thin horny layer; others have a thick operculum with layers of shell material as well as the thin outer layer. A few species have beautifully colored opercula; these are the "cat's eyes" that are used to make jewelry, so called because of their intense green color. The edible conch of the West Indies uses its sharp, horny operculum as an aggressive weapon when it is molested, lashing out at the hand that picks it up.

Some of the sea hares have lost their external shell and have only a small plate embedded in the body as evidence that they once had one. If they are threatened they release a purple ink that seems to be intended to discourage their tormentors.

Perhaps the most remarkable of all defense mechanisms is found in some of the colorful nudibranchs. Nudibranchs have lost all trace of shell, and with it the mantle cavity and the gills, but in some species the upper surface of the body is covered with fleshy cerata (horns), each of which contains a branch of the digestive gland. These nudibranchs feed on coelenterates and manage to digest the prey without setting off the stinging cells. The stinging nematocysts are then positioned in the external cerata, where they are stored in special sacks to be used later by the nudibranch as its own defense mechanism. Some nudibranchs have secondary gills around the anus.

Bivalve mollusks (Pelecypoda) are very numerous and diverse, and some, such as oysters, scallops, and clams, are commercially important as food and in other ways. Their shells are used for building roads and manufacturing cement, and some pearl shells are used for jewelry. The pearl industry is based on the ability of these mollusks to produce a shell layer of great beauty over a foreign object such as a grain of sand lodged in the mantle area. Like the gastropods, the pelecypod mollusks are amazingly diverse, and although they are generally adapted to living in soft bottom environments, there are a few that are able to bore into solid rock or wood if the opportunity arises. The latter ability makes them serious pests, and shipworms, which are actually pelecypod mollusks, do a great deal of damage to ships and piers each year. This particular group of mollusks has lost the radula and has become primarily adapted for straining plankton organisms out of the water currents that pass over the gills. Thus the gills not only function as respiratory organs but also are used for gathering food.

Pelecypods range in size from organisms of a few millimeters to the giant clams of the Indo—Pacific which grow to 3 or 4 feet in length and reach a weight of 400 or 500 pounds. These giants, incidentally, maintain algae in the tissues of their mantles just as the corals do. The algae account for the brilliant colors of the mantle, which varies in color and pattern from individual to individual, with no two exactly alike. Many pelecypods use their muscular foot to kedge themselves along the bottom. Others, such as the thorny oysters, cement their shells to hard bottom. Mussels and smaller species of giant clams attach themselves to the substratum by means of a series of threads called the byssus. Still others, such as scallops, are able to propel themselves by snapping their shells shut and forcing water out through openings to produce a jet-propulsion effect.

The class Cephalopoda includes the squid, the cuttlefish, the octopus, and the chambered nautilus and its relatives. These are much more active than other mollusks and are generally adapted for swimming in midwater, although the octopus is a bottom-dweller. They are characterized by having a number of tentacles which represent a modified foot. The nautilus, a modern remnant of an ancient group, has a coiled external shell, and the Spirula, the beautiful little ram's-horn shell that is so common on tropical beaches, has an internal shell. The shells of squids and cuttlefish are nearly straight and internal, and the octopus has lost the shell entirely. The so-called "shell" of the paper nautilus is actually a shelly egg case.

Squids are often seen by divers. Like other cephalopods they have an extreme ability to change color, and in a newly captured squid, waves of color pulse over the body as it attempts to match its background. Jets of water forced out of the mantle cavity provide rapid propulsion, and both octopus and squid can shoot out clouds of ink to distract predators. Squids and cuttlefish have ten sucker-lined arms, two of which are longer and are called tentacles. These are usually retracted, but when the squid is feeding it stalks until it is close enough to shoot out its tentacles to seize and draw its prey into its mouth. Cephalopods have both a radula and a horny beak, and some have poisonous salivary glands. The beak tears off pieces of the prey, and the radula shreds them into fine pieces.

Cephalopods have separate sexes. Courtship and copulation occur, sperm being transferred to the female by one of the arms. The courtship is quite elaborate. In the spring of the year the cuttlefish enter shallow reef waters. The male approaches the female, and flashes of neon color surge over his body. The male approaches still closer and gently touches the female's forehead, whereupon they unite in an embrace that lasts a few minutes while the eggs are fertilized. The eggs are deposited in clusters or singly under ledges and sometimes are guarded by the female.

The tentacles with their suckers have a rather sinister aspect that has been capitalized on by writers and movie-makers, but most shallow-water cephalopods are small, a foot or so long. There are giant squids with tentacles 35 feet long in the depths of the ocean, but the largest octopus species has arms only 16 feet long and a body about 1 foot long.

Octopods are solitary, but squids often travel in schools of twenty or more individuals, sometimes aligned in rows with military precision. Shallow-water octopus burrows can often be recognized by the "midden," a pile of freshly cleaned snail and clam shells around the entrance. Pursuing the advantage offered by this telltale clue to the gastronomic habits of octopuses, Hawaiians use a cowrie shell with a hook attached to fish for them.

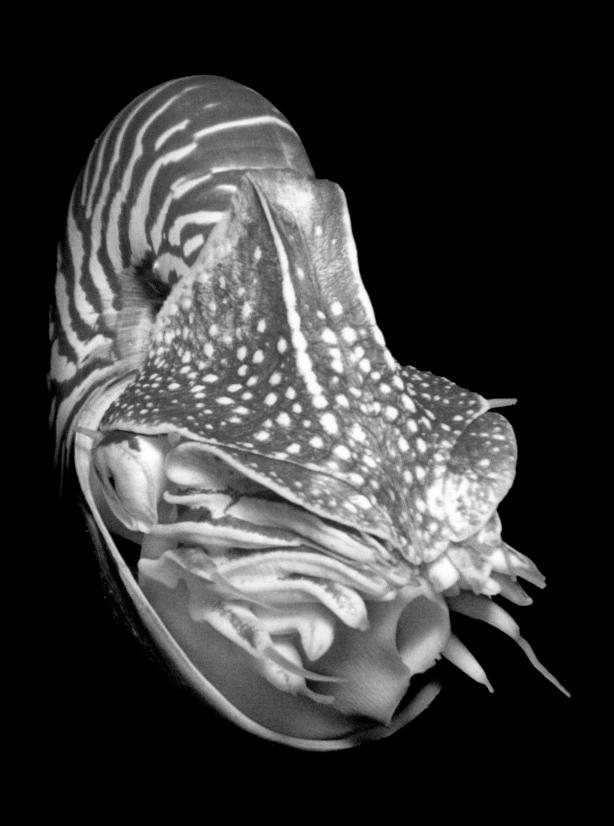

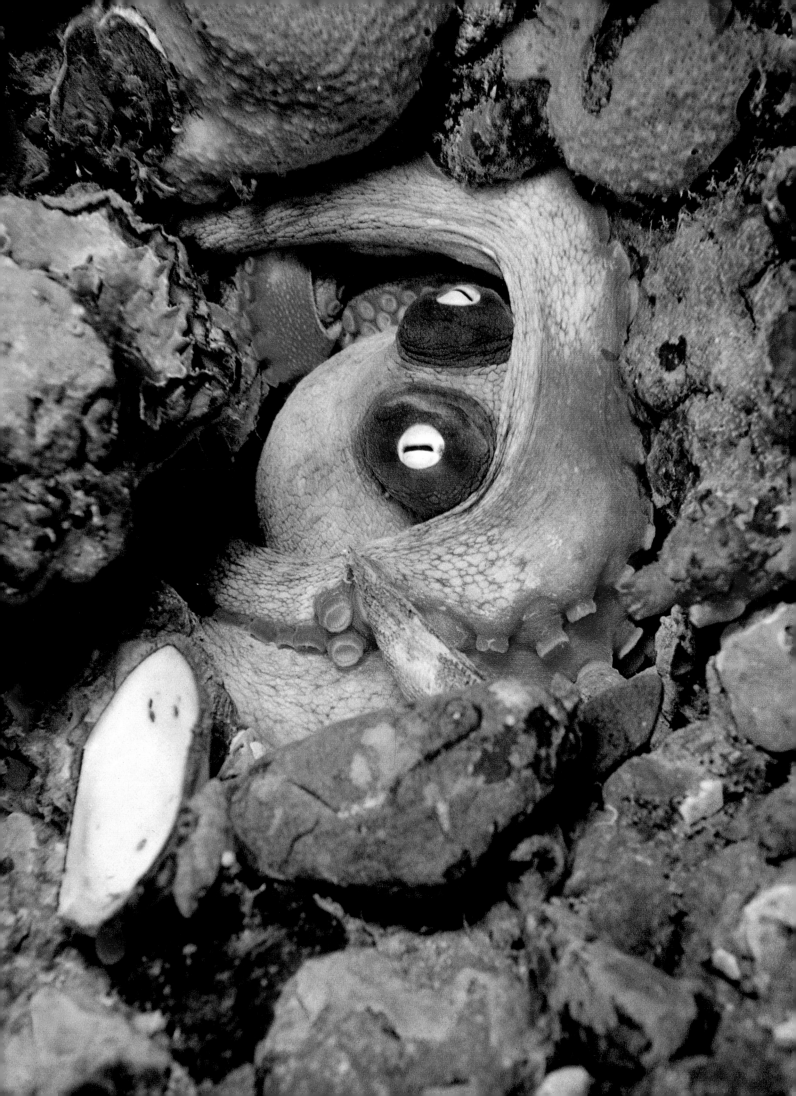

5. CRUSTACEANS

One has only to thread his sailboat into a New England harbor during the lobster season to appreciate the importance of the class Crustacea. For lobsters are members of the Crustacea, a class of the phylum Arthropoda or joint-legged animals. This class includes such important shellfish species as crabs, shrimps, crawfish, lobsters, and a number of forms that fishes and man could well do without—barnacles, parasite copepods, and others of their ilk. Crustaceans make up much of the zooplankton of the sea, and as such they are the chief food of many important species such as herring, menhaden, mackerel, and the young of many other food fishes. Crustaceans occur in marine, brackish, and fresh water, and even in terrestrial habitats, from the depths of the sea to mountain ponds and streams, but they really dominate in the sea and it can be said that most crustaceans are marine.

As typical arthropods, the Crustacea are characterized by a hard external skeleton that is divided into segments separated from one another by flexible joints. Arthropoda is the largest of all of the animal phyla, for it includes the insects, and well over half the animals of the world are insects. From this phenomenal success we can conclude that the basic arthropod design is a good one. It has proved adaptable to almost every environment on this planet—aquatic, terrestrial, and even aerial. Although no animal spends its entire life in the air, insects are strong fliers, and some spiders are able to travel air routes by means of windblown strands of web material.

Arthropod external skeletons offer a maximum of protection at the same time that they provide support for the body and a place of attachment for muscles. They entail certain disadvantages as well, the most important of which are excessive weight and problems with growth. As the arthropod increases in size, the weight of the skeleton increases more rapidly than the power of the muscles, and so there is a practical limit to the size arthropods can achieve. Some fossil species reached a length of 6 feet or more. The king crab spans 12 feet between the tips of its claws, but its body is less than 2 feet across, and this is about the practical limit for this body plan. With the skeleton on the outside, the animal cannot grow effectively, for there is little room for expansion. This problem is solved by periodic molting. The outer skin (shell) is shed and replaced by a new skeleton that forms underneath the old one. This new skin is soft at first, and the animal grows rapidly as its tissues take up water from the environment. After a few hours, as calcium salts are deposited, the new skin hardens and growth ceases until the next molting period. While the skin is still soft, however, it affords little protection and the animal is especially vulnerable to its predators. Certain crabs such as the Jonah crabs choose to mate just after the female molts. Perhaps this practice may serve to protect the female while she is without a hard shell.

Among the arthropods, the insects have dominated the land environments and the crustaceans have taken over the aquatic habitats in the sea and tropical fresh waters. Both groups are conspicuous in temperate fresh waters. There are a number of important groups (subclasses) of crustaceans, which are undoubtedly related, but there are a few features that are present in every crustacean species and can be used for identification.

From detailed studies of crustacean anatomy it is clear that the basic plan of the body is a series of segments, each of which bears a pair of appendages. In the more primitive groups the arrangement of the segments and the form of the appendages are quite variable, but in advanced forms such as the crabs and lobsters there is a consistent number of segments — nineteen. Usually there is a clearly discernible division of the body into the head, thorax, and abdominal regions, each consisting of several segments that have similar modifications for particular functions. Thus, in the lobster the head appendages and segments are modified as the antennae and various jaw parts; the thoracic appendages are the gills and walking legs; and the abdominal appendages are the swimmerets, some of which serve in reproduction, and the fanlike tail. Often the head and thorax are covered by a large structure called the carapace, which is really a greatly expanded dorsal part of the skeleton of one of the head segments. In some groups such as the ostracods and the clam shrimps (Conchostraca) the carapace is divided into a left half and a right half, giving the animal a remarkably clamlike appearance. In other groups the carapace is quite inconspicuous or is modified as a brood pouch in which the eggs are held until they hatch and the young are ready to shift for themselves.

Although the central nervous system is relatively simple, consisting of a ventral nerve trunk and a number of ganglia that serve as a brain to integrate the sensory and motor functions, many crustaceans have elaborate behavior patterns that serve defense, food-gathering, and reproductive functions. These devices make crustaceans a favorite subject of scientists probing the mysteries of animal behavior patterns. Crustaceans often have compound eyes like those of insects. Such eyes are well suited for detecting motion but do not form particularly sharp images. Crustaceans have chemoreceptors for detecting odors in the water (the senses of taste and smell are combined), and sensory "hairs," really small sensory rods, for touch and vibration detection. Although crustaceans do not have ears as such, they can certainly perceive sounds by means of these sensory hairs. They also possess organs called statocysts that respond to gravity and orient their owners so that they can tell the difference between up and down.

Crustaceans reproduce sexually, although some groups such as the freshwater fairy shrimps, tadpole shrimps, and water fleas also produce asexual generations. Usually the asexual generation predominates during seasons with favorable weather conditions, while the sexually produced cross-fertilized eggs are able to withstand dryness or the low temperatures of winter. Many crustaceans have elaborate life cycles and pass through a number of larval stages that are totally unlike the adult. Crab zoea larvae have elongate head spines, and the grotesque phyllosoma larvae of the spiny lobster are so thin and transparent that they are virtually invisible in the water. All crustaceans, however, pass through a nauplius stage, at which time the body is unsegmented and bears three pairs of appendages. Sometimes the nauplius stage is completed before hatching. This is especially true of the higher crustaceans, such as the lobster, which are carried as eggs on the underside of the mother's abdomen and hatch out in an advanced state, often as miniatures of the adults.

Crustaceans, because of their diversity and because some of the species occur in unbelievable numbers, are extremely important in the biological economy of all aquatic habitats, and this importance is in no way related to the size of the individual crustacean. For example, the Cladocera or water fleas, the largest of which is only about ¾ inch long, are the main food of many commercially important fresh-water fishes. In the sea, another group, the Copepoda, are similarly important to those marine fishes that feed on animal plankton. Copepods are also of interest because some of them are parasitic on fishes. A number of these parasite copepods are so modified anatomically that only a study of their life histories can reveal their true identity as relatives of the free-swimming copepods. Somewhat less specialized parasites are the Branchiura or fish lice. These ectoparasites are frequently seen on fishes in aquariums.

Somewhat more familiar to most people are the barnacles. At first glance one might take barnacles for mollusks rather than crustaceans, for the adults are sessile, that is fixed to or burrowing into the substrata, and in this they resemble oysters more than they do shrimps or other "typical" crustaceans. For a long time they were thought to be mollusks, but their life history, including the nauplius stage, as well as their anatomy, leaves no doubt that they are indeed crustaceans. Some barnacles and near relatives of the barnacles are highly specialized parasites, and some of these are parasitic on other crustaceans. One genus even parasitizes parasitic copepods!

Even the most casual visitor to the seashore will encounter a wide selection of crustaceans, many of which are well worth observing for a few minutes or more. As the tide recedes, one can often see fiddler crabs with bright red claws waving off intruders as they scurry about in search of food at the water's edge. Many land crabs dig elaborate burrows and plug the entrances when they are inside. Even while they are digging they remain alert for possible dangers, constantly looking about and retreating at the slightest threat. Many crabs are able to shed a leg easily under attack. A special structure at the second joint of the limb permits the amputation to take place with little damage, and regeneration of the limb starts at once and is completed after a few molts. Better to lose a hand or eye than to lose the whole body.

Most crabs are highly protectively colored. Often they match their background so closely that they escape detection until they move. Some sand-dwelling forms are so adept at digging that they disappear before your eyes with little or no apparent motion. Perhaps one of the most fascinating devices for avoiding detection is that of the decorator crab, which plants pieces of sponge and seaweed on its body, where special crooked hairs hold the camouflage material in place. Members of another group of crabs cut pieces of encrusting sponge to fit their carapaces; and of course the well-known hermit crabs use the shells of snails as residences, changing them whenever they grow out of them. At least two crab species carry anemones in their claws and use them as weapons for defense and for stunning small organisms for food.

The mantis shrimps (Stomatopoda) are spectacular crustaceans often encountered in shallow seas. As their name implies, they have large abdomens and large claws that resemble those of the praying mantis insects. The abdominal appendages are modified as respiratory organs, and the tail segments are usually sculptured and are sometimes used to block the entrances to the burrows they build. Some stomatopods get to be a foot or more in length and are aggressive carnivores, feeding on almost anything that happens by, including others of their own species. Stomatopoda larvae are extremely transparent and have long spines projecting from their carapaces. Mantis shrimps should be treated with respect, for their huge spiny claws can give a painful stab to anyone who tries to handle them.

From an economic standpoint, shrimps are perhaps the crustaceans most directly beneficial to man. The shrimp fishery is one of the largest in the world. New fishing grounds are still being discovered, and as better techniques for harvesting these delectable crustaceans are developed, the industry continues to grow. Efforts are being made to find ways of raising shrimps and other crustaceans under controlled conditions. There is little doubt that in the future much of our food supply will come from the sea, and it is certain that crustaceans will be among the most important organisms cultivated and will provide a substantial part of the world's protein supply in the foreseeable future.

6. ECHINODERMS

Although the name Echinodermata means "spiny skin," not all members of the phylum bear spines: some sea cucumbers are soft and wormlike or resemble large sausages. Perhaps the most familiar members of this fascinating phylum are the starfish, yet no one who has gone diving in tropical waters will be likely to forget the sharp spines of sea urchins. But don't look for echinoderms in fresh water, for they are exclusively salt-water-dwellers, perhaps because they have no excretory system to rid themselves of excess water that would enter their bodies in fresh-water environments. Echinoderms are radially symmetrical—that is, the parts such as arms extend outward from the central part of the body—but this does not mean that they are closely related to other radially symmetrical animals such as the corals and sponges. There is abundant fossil evidence of the larval stages to show that the echinoderms are derived from bilateral ancestors, and their symmetry is a secondary adaptation to a bottom-dwelling existence. Echinoderms have a magic number of five. Their parts occur in fives or multiples of five. The starfish have five arms, the sand dollars have five grooves or holes in their shells, and the sea cucumbers have five double rows of sucker-like tube feet. The living Echinodermata are divided into five classes: the Asteroidea, or starfish; the Echinoidea, which includes the sea urchins and sand dollars; the Ophiuroidea, encompassing the brittle stars and the basket stars; Holothuroidea, which embraces the holothurians or sea cucumbers; and the Crinoidea, including the feather stars and sea lillies.

Echinoderms have a skeleton that is made up of small calcareous plates. These plates are developed to different degrees in the different groups. In starfish, for example, they are irregularly shaped nodules, but in the sea urchins they are exquisitely sculptured and shaped so that they fit together to make a solid shell that is familiar to beachcombers as a "sea egg." In the sea cucumbers the plates are microscopic in size, but some look like exquisitely sculptured buttons, tables, and wheels. These plates are used extensively for the classification of the species of sea cucumbers.

Echinoderms have a unique circulatory system called the water vascular system. This is a system of tubes that carries sea water to all parts of the body. Its pressure is used for maintaining the body shape and operating a series of sucker-like tube feet. These tube feet, which sometimes occur in pairs, thus in multiples of ten rows, are one of the chief means of locomotion and aid in holding the animals to the bottom. The water vascular system is also used for breathing, so in a sense it is a single highly efficient system for respiration, feeding, and locomotion.

Echinoderms vary widely in their feeding habits. Starfish are canivores, feeding on mollusks, crustaceans, polychaete worms, corals, and other echinoderms. Off Maine, a diver has observed a starfish and several sea urchins feeding on a sand dollar. Every beginning biology student knows how the starfish surrounds a victim such as a clam by using its everted stomach extended outside its body to engulf the meal. Other echinoderms, however, swallow their prey whole or rasp off pieces and swallow them.

Sea urchins have a remarkable set of five jaws that together form a complicated structure somewhat resembling an old-time candle lantern, and so it has been given the whimsical name of Aristotle's lantern. The lantern is an efficient chewing device, but it is quite slow. Most sea urchins feed on almost any organic matter, but some urchins and sand dollars burrow just below the surface of the bottom and feed mainly on detritus and other small particles in the sand or mud. Brittle stars and basket stars, which can be distinguished from starfish because their arms are sharply set off from the central body, feed on detritus and small living and dead animals. Their digestive tract is extremely simple, with no intestine and no anus. Indigestible materials are expelled through the mouth. Basket stars, brittle stars, and feather stars all trap plankton and detritus with their arms and carry them along grooves toward the mouth. During the daylight hours the basket stars hide, curled up, under coral ledges. At night they come out and spread their delicate branched arms to gather the countless plankton drifting with the tide. Even more interesting are certain sea cucumbers. These stout-bodied elongate animals trap drifting plankton with their tentacles. Then they stick their bushy tentacles, one after another, into their mouths and clean off the trapped particles, just as we might lick gravy off our fingers. Other sea cucumbers bulldoze the bottom and pass large amounts of sand through their digestive tract, thus acting as important scavengers in many tropical areas.

Most echinoderms have separate sexes, although some are protandrous—they function first as males, then become females. They also have great powers of regeneration. If an arm is lost, they are able to grow a replacement with little trouble. Some species carry this process a step further and reproduce asexually by dividing in two. Usually the eggs and sperms are released into sea water and fertilization is external, but a number of echinoderms of all classes, especially the cold-water forms present in the arctic or antarctic, brood their young in some sort of pouch. After spending the planktonic larval stages in the open sea, the young settle to the bottom and take up their adult way of life. Most echinoderms are stay-at-homes and some of the sea lilies become permanently attached to the bottom, but others move about freely and some of the feather stars can swim by moving arms alternately up and down, like the oars of a Roman galley. Although echinoderms are not the most active animals in the sea, they nevertheless have a lot to offer the diver who takes the time to observe them. They are highly specialized and have many unique anatomical features, and they also attract attention because there is evidence suggesting that they are more closely related to the chordates than is any other group of invertebrate animals.

When you enter the water on your way out to a reef, you will not have long to wait before encountering your first echinoderms. These will almost surely be spiny sea urchins living near any rocks along the shore. Here you will find the long-spined *Diadema* tucked into crevices in the rock; and in some of these areas you will find a dark red, short-spined form living in channels carved in the rocky surface. These channels are about as deep and as wide as the urchins and several times as long as their diameter measured to the tips of the spines. There is little doubt that the channels were carved by the urchins themselves, but just how it was done is not certain. At any rate the urchins in these channels are well protected, as you will find if you try to dislodge one. In fact, it will probably be necessary to break the urchin's spines and maybe the central test (or shell) as well in order to remove it.

There is one species of urchin, quite common in the Hawaiian Islands, that is adapted for life in the intertidal zone in a different way. Instead of hiding in crevices, it lives out on the rock surface along with chitons and some snails. It can do this because its spines are reduced to short plates. Those on top are short and hexagonal, those on the edges are paddle-like, presumably to deflect waves and enable the urchin to hold on.

Farther out, in beds of sea grass and algae, one is likely to find urchins with short white spines, to which pieces of dead coral, rock, shell, and plant material are held in place by the urchin's tube feet as a simple kind of camouflage for an animal that is otherwise quite vulnerable to predation. We also find that certain sea cucumbers resort to the same kind of camouflage.

Large sea cucumbers are usually prominent in shallow-water areas, but there are many small burrowing or hiding forms that are never seen by the diver. There are even some peculiar sea cucumbers in very deep water. These animals have been able to adapt to almost the full range of marine bottom environments. Sea cucumbers, at least certain species, are harvested in the East Indies for use in some Chinese dishes. For many years the trepang or bêche-de-mer fishery was one of the most important industries in certain areas of the Great Barrier Reef. The cucumbers are boiled and then dried in the sun. The muscular body wall is the part that is used.

Some sea cucumbers hide by burrowing in the sand; others with slimy or warty, often colorful, surfaces don't bother to try to conceal themselves; they have no predators to fear. Many of these species are quite poisonous— so much so that an injured sea cucumber will kill any fish placed in the same aquarium tank with it. Native fishermen in many places know and use this toxic property of the sea cucumbers. Whenever they find a fish which they can't dislodge from a hole, they grab a sea cucumber and rub it on the rocks around the hole. Toxic substances from the injured sea cucumber then irritate the fish and drive it out of its hole into a waiting net. Scientists have studied these substances intensively, and one of the materials has shown some promise as an inhibitor of the growth of cancer cells.

Sea cucumbers have still another defense mechanism, although it is present in only certain species. This is a specialized organ that produces quantities of an extremely sticky material that is released in threads the size and color of spaghetti. The threads are not actually sticky until they are ejected and come in contact with the sea water, which serves as a catalyst. Once the threads are activated, they are so adhesive that if you get them on your arm it will be necessary to shave the hair off in order to remove the mess. In some areas the natives use this material as a bandage if they chance to cut themselves while out on a reef. A sea cucumber of the right species can be used like a tube of glue to bind up the wound.

The final defense of many sea cucumbers is self-evisceration. If they are attacked, or if conditions become unfavorable, they expel their internal organs—digestive tube, respiratory trees, reproductive organs, and all. Presumably this satisfies or discourages predators, although it is hard to see how. At any rate, the sea cucumber then goes on to grow a new set of organs.

This may take several months, but eventually the job gets done and the sea cucumber is as good as new.

Although most sea cucumbers are slow-moving, there is one group, called the synaptids, whose members are capable of fairly rapid movement. Some of these reach a length of four or five feet, and they look almost like snakes as they amble along in search of food. If you try to lift one of these organisms out of the water it will collapse like a wet silk stocking. The body walls of these animals are so thin that they have lost their respiratory trees in favor of direct absorption of oxygen from surrounding sea water. Perhaps one of the most fascinating features of these creatures is the form of the spicules in their skin. These are exquisitely formed anchors and wheels, and under the microscope one can easily see how the anchors hold on to whatever the animal touches. All in all, the placid sea cucumbers are well worth the attention of the diver who wants to become acquainted with them.

Starfish are also fascinating to watch. Again, they are not rapid, but they get there in their own good time. The variety of shapes and colors is seemingly endless. The large West Indian sea star *Oreaster* shows a subtle gradation of yellows, browns, red-orange, or combinations. On the reefs of the Indo-Pacific one occasionally encounters a brilliant cerulean-blue starfish that wraps its smooth, slender arms around and through crevices in the coral. Certain burrowing forms make their way into sand, leaving star-shaped tracks to mark the point of entry. One of the more fascinating shapes is that of the so-called pillow star, which has arms so short that it is reduced to a pentagon. Alive, it resembles a slightly angular grapefruit. The five grooves radiating from the mouth show that it is indeed a sea star.

More active and mobile than the sea stars are the brittle stars and basket stars. Brittle stars are common everywhere around rocky reefs. They average an inch or so across the disk, with arms three to five inches long. Some are smooth, some spiny. When a brittle star is harassed, one or more arms may break off and crawl away as if they were independent caterpillars. These stars also range in color from gray to dark purple to bright orange.

Sea lilies and feather stars are relics of a very ancient group of echinoderms. Stalks of sea lilies are extremely common in Devonian rocks, and I have heard them described as "fossilized corn cobs." Today stalked sea lilies are confined to deep and cold waters, but about two hundred fifty feather star species live in shallower, warmer waters. Many are extremely beautiful, displaying every color of the rainbow—and then some.

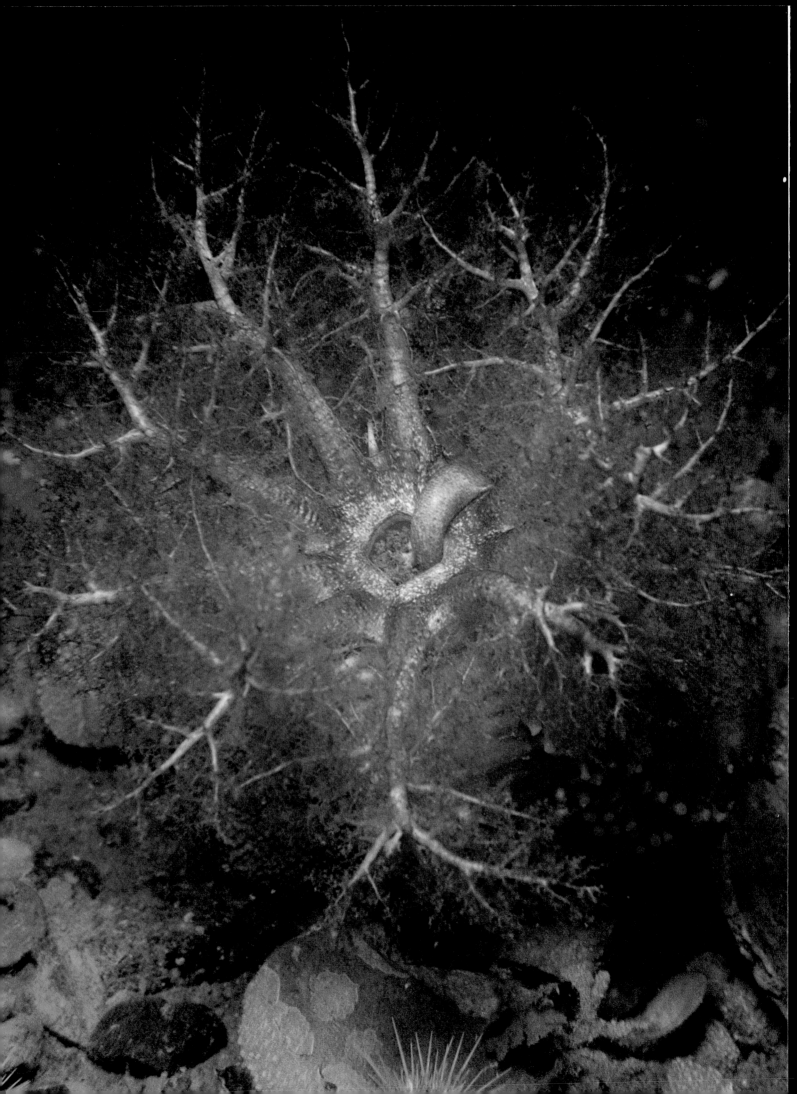

7. FISHES

The only backboned animals that are really abundant in the sea are fishes. An occasional turtle, a sea snake, or a porpoise may now and then be encountered by the diver who spends a lot of time under water, but these are rare compared to the thousands of species of fishes. At best, the other vertebrates, being air-breathers, must periodically return to the surface to breathe, whereas fishes have no such restriction.

Many fishes, especially those that live around coral reefs, are brightly colored, but others are quite drab, and some resemble the rocky bottom, complete with fleshy skin tabs that simulate attached algae. Usually those fishes (such as the scorpion fish) that utilize protective coloration remain on the bottom and swim very little. They depend for protection on not being noticed. They also have the advantage of being able to approach other fish closely, to catch a meal.

In tropical areas a diver under water is sure to be impressed by the profusion of fishes. In some places they occur in such numbers that one can scarcely see the bottom. Nevertheless, it is numbers of species rather than numbers of individuals that make the fish communities some of the most fascinating in the world. It is quite usual to find fifty species living together in a coral stack no larger than an office desk, and a hundred or more may coexist along a hundred-foot stretch of rocky shoreline. The riddle of how these species manage to share the available resources of food, shelter, and living space in a state of dynamic stability is one of the most challenging problems in tropical marine biology today.

Fishes that live around coral reefs do so because they find in that environment all the things they require. Food, shelter, nesting sites, and a reasonably stable chemical environment make a reef ideally suited for many kinds of fishes. While some of the species found around reefs are large, reaching weights of a hundred pounds or more, most reef species are less than a foot long, and the maximum size reached by some gobies is less than an inch and a half. Many coral-reef fishes have distinctive shapes: the trumpet fish is elongate with tiny fins and drawn-out face, while the butterfly fish has a short, deep, flat body that gives it extreme maneuverability around the corals. Gobies and clingfish have their pelvic fins united into a sucker with which they fasten themselves onto corals or the bottom. Eels and some other elongate fishes are able to work their way into holes in the rock or to make burrows in the loose bottom sediments around the reef structures.

The general appearance of a particular fish provides many clues to its habits. Surface-dwelling forms such as the needlefish and halfbeaks are slender and are colored so as to be almost invisible near the surface. Barracudas, ceros, sardines, and jacks are midwater fishes that feed from near the bottom to near the surface. Their bodies are spindle-shaped or compressed, and their deeply forked tails provide power for the fast swimming needed to catch food and outdistance enemies. Bottom-dwellers have more varied shapes and are either brightly colored or colored to match the bottom. The utilization of different microhabitats and living habits enables these fishes to live side by side with a minimum of competition for space.

One very important activity is the never-ending hunt for food. To minimize competition, different fishes feed on different foods at different times, and their feeding cycles are timed to their prey. Parrot fish feed actively during the day, but squirrel fish are nocturnal feeders. Their food habits are also reflected in the anatomy of their mouths and digestive systems. We can, for example, recognize plankton-feeders by their small mouths. Those that feed on plant plankton have numerous long gill rakers on the gill arches that act together to form a filter for straining plankton organisms from the water that passes over their gills. Fishes that feed on larger plants have their teeth specialized for cutting and nibbling and the longer intestinal canals needed to digest plant matter. They also have some sort of mechanism for grinding their food because most of them lack sufficient enzymes for digesting the cellulose walls of plant cells. Therefore the cells must be crushed before they can be digested. In the parrot fish this is done by pharyngeal teeth that form a grinding mill, but in some surgeon fish and in the mullets the grinding is done in a muscular gizzard-like stomach. Carnivorous fishes such as sharks and barracudas tend to have large mouths. Mollusk-feeders like the porgies usually have molar-like teeth for crushing the hard shells of their prey, and fish-eaters generally have sharp teeth for grasping and holding their food.

Many fishes have mouths that are specialized for particular feeding habits. The longnose butterfly fish, for example, whose mouth is at the end of a long snout, can reach into crevices for small crustaceans, and the mouths of certain wrasses and mojarras can be extended to capture prey from its hiding places. The goatfish or surmullet has a pair of prehensile barbels that detect food and animals hidden in the sediments as the fish cruises along, skimming the bottom. Once the food is located, the goatfish stops and goes after it, often burying its head up to the eyes as it pursues its sand-dwelling prey. One of the best-known reef-feeding specializations is that of the parrot fish. The teeth of these fish have coalesced into powerful beaks (hence the common name), and the fish bite off chunks of plants and living corals, digest the organic matter, and pass the ground-up sediments as a fine silt. Trigger fish have powerful jaws as well as tough hides that are useful for attacking spiny-shelled sea urchins and mollusks. The food habits of most fishes change as the fish grows, so the juveniles may have a diet quite different from that of the adults. Juvenile groupers feed mostly on crustaceans and become progressively more piscivorous as they get larger.

Sometimes it is difficult to decide whether a particular specialization serves primarily for food-gathering or as a protective mechanism, because many modifications serve both purposes. In truth, the functions of feeding and defense are parts of the same phenomenon, for every predator has its enemies as well as its prey. Concealing colors, for example, enable a fish to stalk its prey without being detected and at the same time conceal it from its own predators. This is well illustrated by the predaceous little sargassum fish, which bears a striking resemblance to the seaweed in which it lives. Body form can also be concealing. Many scorpion fish are exceptionally well camouflaged by their shape; to the extent that on one occasion I tried to pick up the fin of a scorpion fish, thinking it was the ribbed shell of a Spanish oyster. Concealment is often combined with other modifications. The fishing frogs, diminutive relatives of the northern angler fish, have a modified dorsal fin spine with a fleshy tip that serves as a fishing lure. Against its normal background the fish is almost invisible, but the lure wriggles on the end of the pole like a small worm. Woe to any fish that tries to eat the "worm," however, for as soon as one approaches close enough the angler fish takes a quick gulp and "inhales" a meal.

Many fishes have a protective armor plating that renders them almost invulnerable to most predators. The cowfish and the boxfish are encased in bony capsules of modified scales and produce a toxic slime as well. Sea horses and pipefish are protected by bony rings, also modified scales. Sharp and venomous spines constitute another successful line of defense. Most perchlike fishes have spines in the dorsal, anal, and pelvic fins, and the angel- fishes add a stout spine on each cheekbone to this weaponry. The squirrel fish even has all these protections plus spines on its scales, and fishermen know to treat any squirrel fish they catch with due respect. The porcupine fish, a large puffer, has scales that are shaped like jackstraws. When in danger, it can inflate itself with air or water until it becomes nearly spheri- cal, and with all its spines erect it looks like a prickly basketball. It also locks itself into crevices by filling with water, jamming itself in place so tightly that it cannot be pulled out. The trigger fish gets its name from its dorsal-spine mechanisms. A cam on the second spine locks the first spine in the erect position and holds it there until the second spine is depressed. It also wedges itself into reefs and erects its spine to defy its enemies.

No defense mechanism can be successful unless there is a proper be- havioral pattern to go with it. It would do no good for a fish to resemble an algae-covered rock if it spent its time in midwater or resting on sand. But sometimes the element of surprise comes in. Juvenile spadefish, along with several other species, are jet black, but they spend their time on white sand, where they are extremely conspicuous. However, on sandy bottoms there are usually bits of black debris such as mangrove seeds or leaves, and the little spadefish so closely resemble these plant parts that they are almost impossible to see.

Fishes that are fast swimmers or able to maneuver quickly often rely upon their ability to outrun their predators as their first line of defense. This flight reaction is carried to extremes in the open-water flying fish which leave the water and glide long distances when pursued. Since fishes in the water can- not see through the surface layer except for a very limited area nearly straight up, the predator is unable to keep track of a flying fish. At the opposite ex- treme, some wrasses dive straight down into the sand with the speed of a magician's rabbit disappearing into a hat: now you see him, now you don't.

Eventually fishes, like all other animals, must reproduce, and their spawn- ing habits are almost endlessly varied, although we know very little about the reproductive life of most of them. Many have elaborate courtship rituals, and it is to be supposed that brilliant colors serve for species- or sex-rec- ognition. This latter is of course especially likely in those species in which the male is colored differently from the female. A number of fishes such as the sergeant majors and the damsel fish guard their eggs until hatching. The eggs are fastened to a rocky surface, and the father guards the eggs by driving away other fishes that try to eat them. He also cares for the eggs by fanning them to keep sediment from settling on them, as an unchecked accumulation would eventually smother the defenseless embryos. Even greater parental care is exhibited by the males of some cardinal fish and jawfish, which carry their eggs in their mouths until the eggs hatch. Many other fishes, however, exhibit no parental care. Their eggs float free in the ocean as part of the plankton. Such eggs are buoyed by a prominent oil droplet that acts as a float. After hatching, the young of many reef species remain as plankton until the ocean currents carry them to a suitable habitat. During this planktonic larval stage they bear little resemblance to the adults they will become. They are transparent and often have long trailing spines that apparently serve to retard their sinking rate. When they reach a proper environment they transform quite rapidly, often shrinking in the process. Their pigment develops and they assume the general shape of the adult stage, although they are still very small. Larvae that ultimately fail to find an acceptable habitat eventually perish, but there is some leeway. Amazingly, if they don't find a suitable environment at exactly the proper time they can remain larvae for a while longer and continue drifting. As soon as the proper environment is found, they immediately begin their transformation.

8. CLEANING SYMBIOSIS

Predator-prey relationships are not the only interspecies relationships in natural communities. Quite a few have developed cooperative arrangements whereby two different species live together in a way that is decidedly beneficial to one or both without seriously inconveniencing either. We have already mentioned the arrangement in corals and giant clams where algae live within the tissues of the animals. The algae use the animal's carbon dioxide wastes and in turn supply oxygen and some nutrients to their host. We have also seen how shrimps, worms, and fishes have set up residence in sponges. While it has not been demonstrated that this is of any benefit to the sponge, it might be postulated that the movements of the residents may enhance the water currents through the sponge, and perhaps their waste materials are used by the sponges. At any rate, it appears that the presence of the tenants does little harm to the sponges, for the largest and healthiest-appearing sponges are always teeming with residents. Biologists at first had the temerity to decide which relationships benefit both and which benefit only one of the species involved, and such terms as commensalism and mutualism were coined to suggest how the rewards were divided. Upon a closer look, however, it seems well to be cautious and make such judgments only after extensive and carefully controlled experiments.

Often interactions are widely recognized without our realizing their full importance. The so-called cleaning symbiosis, which we now know a great deal about, was once considered to be a mere biological curiosity. As more and more divers began to explore the ocean, it became apparent that the phenomenon is widespread and that a number of fishes and shrimps make a practice of cleaning ectoparasites and other fouling organisms off fishes. Subsequent experiments have led to an understanding of the real significance of the cleaning phenomenon. Cleaning symbiosis in the sea has parallels on land: the Egyptian plover cleans crocodiles, the cattle egret services cattle, and the tickbird grooms the rhinoceros. There are even some parasite-removers that clean parasitic mites from larger insects, and crabs that remove mites from marine iguanas in the Galápagos Islands. Thus it seems that the need for parasite-removers is universal and has been met satisfactorily in different groups of animals living in different environments.

Cleaner fishes vary considerably in their habits, but observations and experiments have shown that there are some similarities that characterize the relationships. The cleaners generally establish certain territories or "cleaning stations" to which other fishes come, sometimes from long distances, to be cleaned. Other cleaners, however, can be likened to doctors on house calls; they travel with their patients and perform their services as opportunities arise.

Cleaners tend to be distinctively colored, with bright yellow or blue stripes, so that they can be recognized. They also tend to signal the client by distinctive hopping-swimming movements. Cleaner shrimps have white antennae, which they wave to attract attention to their presence. For its part, the fish that needs cleaning responds to the signals of the cleaner by approaching and then stopping and offering its head or body to the cleaner. Cleaners are generally immune to being eaten by the fishes they service. Amazingly, they often swim right into the mouth or gill covers, and after a moment or two come back out even though the species being cleaned may be a carnivore that feeds on prey of the same general size as the cleaner. Evidently there is a warning system by means of which the patient tells the cleaner to leave. One scientist has reported watching a small wrasse cleaning a moray eel in the waters off Bora Bora. The four-foot-long moray opened its mouth, and the wrasse entered. After about eight seconds the eel signaled by moving its head to one side, whereupon the wrasse "fairly flew out of the mouth and the [eel's] awesome jaws snapped shut!"

In the Pacific four species of wrasses are cleaners, and one might be tempted to suggest that this is because most wrasses are brightly colored and can be easily recognized. In the Atlantic the cleaning habit exists among juvenile angelfish and brightly colored gobies as well as some of the local wrasses. Large angelfish are drab-colored and feed largely on sponges, but the juveniles that are cleaners are boldly marked with golden bars on a nearly black background. Clearly the color-pattern transformation is correlated with the change in feeding habits. Cleaners work on a variety of fishes, from surgeon fish and groupers to morays and barracudas, and even the schools of jacks and the giant sunfish of the open waters come into the shallows to be cleaned. In the open sea the remora or shark-sucker seems to service the sharks, for shark parasites have been found in the stomachs of remoras.

One scientist postulated that the presence of large numbers of fishes over the continental shelf may be due to the need for being cleaned periodically. Some cleaners are agreeable to cleaning anything that comes their way, and will even work on a human diver if he presents himself at their station. Most cleaners feed on other organisms, but they also eat the parasites that they remove from fishes. However, some of the tropical cleaners are kept busy doctoring much of the time and must rate as nearly full-time parasite-eaters. In addition, they remove fungus, plant growth, and dead tissue around any wounds.

Cleaners seem to attract fishes from a considerable distance. Often the stations are established in well-marked locations such as large coral heads or stacks, wrecks, and kelp beds. Sometimes several cleaners work together at a station and the fishes to be cleaned form an aggregation and wait their turn. Conrad Limbaugh in an article in *Scientific American* for August 1961 recalled that he once observed a cleaning station during a six-hour period and counted some three hundred fishes being cleaned. He noted that certain individuals that could be identified by particular marks returned day after day.

Limbaugh also pointed out that cleaners are extremely important in the economy of the reef and he performed some experiments to evaluate their role. He selected two small reef patches and removed all the known parasite-pickers, including shrimps. Within a few days fishes began to disappear from these reefs, and many of those that remained had frayed fins and ulcerated sores on their bodies. Subsequent experiments in another part of the world gave somewhat less clear-cut results. It appears that local conditions may influence the importance of cleaning symbiosis.

As has been mentioned, fishes are not the only parasite-pickers. Aside from the nearly thirty species of fishes, at least six species of shrimps also indulge in the practice. Like the fishes, the shrimps are also brightly colored, yet they seem to signal their clients mostly by waving their antennae in a characteristic conspicuous manner. It is not uncommon to see these shrimps climbing around in the mouth of a moray eel as well as on the gill covers of other fishes. A number of the shrimps, including the Pederson shrimp, live in association with sea anemones, either clinging to them or occupying the same hole. Some shrimps manage to clean fishes without leaving their burrows; others actually come out and climb aboard the patient. Most cleaning shrimps are strongly territorial and probably don't move more than a few centimeters during their lifetime. Evidently the relationship is still evolving in some species, for a California shrimp, *Hippolysmata californica,* has only an imperfect cleaning relationship. Its color is rather undistinctive, it occurs in groups, and it is not known to display to attract fishes for cleaning. However, although it will clean fishes that occupy the same holes in which it seeks shelter, it will also feed on small fishes. It frequently is eaten by moray eels, so apparently it does not have the immunity that the full-time cleaners have.

There seem to be impostors in many fields, and the cleaning business is no exception. In the Indo–Pacific area there lives a blenny that has a striking resemblance to one of the cleaner-wrasses, *Labroides dimidiatus.* But whereas the wrasse feeds on parasites, the blenny *Aspidontus* feeds on the patient. To be able to accomplish this, the blenny has evolved as a mimic of the cleaner. The mimic is colored very much like the *Labroides* and resembles *Labroides* in size and shape as well. Even more remarkable, it swims in the characteristic manner of *Labroides.* There is no doubt that this species mimics the cleaner wrasse and takes advantage of its immunity from predation to dart in and tear chunks of skin from the fishes that offer themselves for cleaning. However, larger fishes with some experience, no doubt gained the hard way, seem to learn the difference between cleaner and mimic and avoid the blenny. The mimicry of the blenny is so complete that the juveniles assume adult colors at about the same size as do the cleaners, and like any good mimic they are less abundant than the model they represent. If this were not the case, fishes that too often encountered the mimic would also avoid the model, and none would benefit. A similar case of mimicry occurs in the West Indies, where the wrasse *Thalassoma bifasciatum* is mimicked by the blenny *Hemiemblemaria simulans.* In this case the blenny is not known to bite the skin of larger fishes that come to be cleaned, but it does eat small fishes. John E. Randall has suggested that the mimic may benefit from resemblance to the cleaner which does not eat fishes. Consequently the blenny mimic can approach its prey without alarming them.

From the foregoing it is obvious that cleaning symbiosis is an extremely important interspecies relationship. Irenaus Eibl-Eibesfeldt has theorized that the peculiar swimming movements of the cleaner wrasses evolved from an initial desire to flee on the part of the tiny cleaner, which at the same time still wants to approach the potential predator. The swimming motion is a cross between fleeing and approaching. That this elaborate ritual behavior evolved over a great period of time is evidenced by the fact that the blenny mimic also required ages to attain its status. The fact that a patient is willing to risk periodic attack from the mimic for the continued benefits of the real cleaner demonstrates the tremendous importance that the parasite-pickers have in the fish community.

There is still much to be learned about cleaning symbiosis, but it is now apparent that it is extremely important in the economy of the fish populations of coral-reef areas. If, as seems likely, there are other important but subtle interspecies relationships that have gone unnoticed or unappreciated, the observant diver can gain significant clues to the reef biology merely by keeping alert to such phenomena.

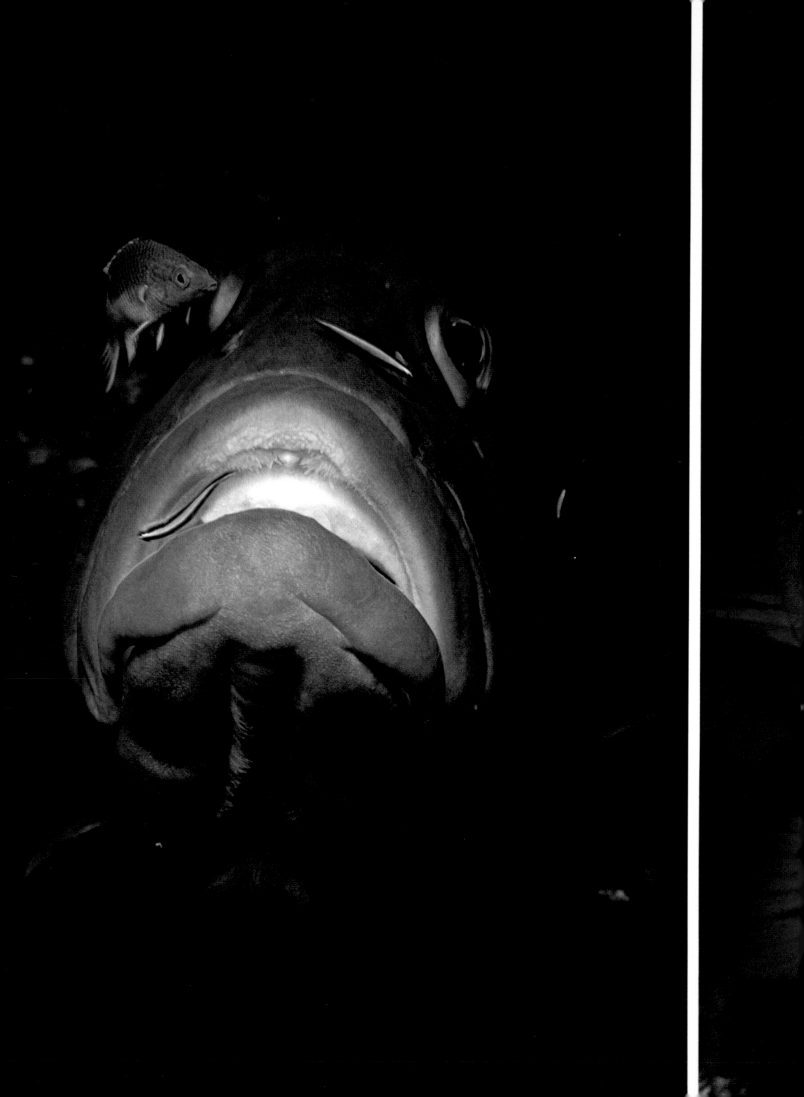

9. DANGEROUS MARINE ANIMALS

Man does not belong in the sea. He is not designed for breathing or for catching his food in water and, even using his best Olympic style, he is not designed for swimming through the water. His impervious skin with its sweat glands is not made for long-term immersion, and the numerous air spaces in the lungs and sinuses and ears are vulnerable to the effects of pressure. His sense organs are strictly designed for aerial work; his keen sense of sight is of little use in the water, and his hearing mechanism is designed only for picking up airborne transmissions. All in all, man in the sea is, generally speaking, as far out of his element as is man in space.

On the other hand, animals that belong in the sea have managed to survive there because they are adapted to it. For them the sea is not a hostile environment but a harmonious one. They are attuned to its temperature, movement, pressure, and light conditions as we are attuned to the soft breezes of a summer day. But there is conflict in the sea, conflict that arises out of the simple fact that animals, being unable to synthesize their own foods as plants do, must feed on plants or on other animals. This being the case, there are innumerable predatory-prey relationships, and every time a carnivore eats, a prey suffers its ultimate fate. This predator-prey interplay has led to a fascinating array of interacting adaptations. In evolution the success of a species is determined by its reproductive success—how many offspring survive, generation after generation. Those individuals that are most successful at avoiding their predators are most likely to survive to produce the most offspring, provided they don't starve to death in the process. As the prey evolves more efficient defense mechanisms, the predators must evolve more efficient food-capturing mechanisms, and the ultimate resulting race is reminiscent of a Madison Avenue advertising war. All this is complicated by the fact that each animal is both predator and prey.

When man enters this battleground he finds himself in the middle, and as a stranger he is in many ways a victim of the predator adaptations of some animals—sharks, for instance—and the recipient of defense mechanisms, in the form of venoms and spines, of other animals such as the lion fish.

Dangerous marine organisms can, for convenience, be grouped into those that cause mechanical damage and those that are toxic in one way or another. The latter group includes venomous species that by their bites or stings cause discomfort from poisonous substances, and toxic species that cause distress when they are ingested. Relatively few animals in the sea are large enough to attempt seriously to eat a man, and the only really severe danger comes from sharks, of which some seven or eight species out of over two hundred fifty are really man-eaters. Killer whales are potentially able to attack men, but they are not abundant near shore and can generally be discounted. Barracudas occasionally attack man in the water, but the authenticated records of barracuda attack are rare. Sharks and shark attacks have recently been the subject of a great deal of study, and we know more than we did before, but we still don't know exactly what causes sharks to attack, nor do we know what to do about them. Since sharks feed mainly by smell and sound, it follows that if blood from a wounded person can be contained, the danger of shark attack will be reduced. To this end plastic bags have been devised for downed aviators to isolate them from the surrounding waters. These have the added advantage of keeping the person warm, since the water in the bag soon warms to near the person's body temperature and forms an insulating layer. Certain colors are less likely to attract sharks, and they can be a factor in the effectiveness of such devices. Earlier attempts to find a really good shark-repellent have met with indifferent success, although sharks do tend to avoid black dye clouds in the water. A major factor in the shark-attack problem is the fact that sharks go into a "feeding frenzy" when they find food and attack repeatedly even after they themselves have been seriously wounded. Once their frenzy behavior pattern starts, there is no known way of turning it off.

Toxic substances vary in their effects, depending on how much poison the victim receives and his individual sensitivity to the substance. Just as some people are almost immune to poison ivy and others are seriously affected by it, the effects of poisons administered by marine organisms vary with the individual and his physiological condition.

Many organisms use toxins for defense and also for subduing their prey. The coelenterates such as jellyfish are distinguished by their stinging nematocysts. The Portuguese man-of-war, with its very long tentacles, and the sea wasp jellyfish can deliver enough venom to paralyze or even kill a human being. Other jellyfish merely produce a stinging sensation or a mild rash. Fire coral, the stony hydrozoan that resembles the true corals, can cause a painful burning sensation. Fire coral is particularly annoying because it often encrusts the dead skeletons of sea fans and may not be noticed until a diver rubs against it.

Poisonous coneshells, which have radular teeth that act as tiny harpoons, can cause discomfort and in a few hours even death in human beings. It is thought that certain coneshells feed on invertebrates and that their venom is designed to subdue their prey. Other coneshells feed on vertebrates (fishes), and consequently their venom affects backboned animals. Since man is a vertebrate, it seems apparent that only those coneshells that feed on vertebrates are dangerous to man. This is fascinating to contemplate, but it is safer to regard all coneshells as potentially dangerous.

Another invertebrate that should be treated with respect is the crown-of-thorns starfish. It has many long venomous spines, any one of which can cause a painful puncture wound. Consequently anyone falling or stepping on one with bare feet could be in serious danger. Another echinoderm that can cause a painful wound is the long-spined sea urchin of the genus *Diadema.* Our experience has been that the effect is uncomfortable but short-lived. In this species the spines are extremely fragile and break off in the flesh but will be absorbed by the body in a few days. Stouter spines from other sea urchins may not be as painful upon entry but will remain embedded and fester until they are removed or work their way out.

Divers should also give wide berth to the bristle worms of the genus *Hermodice.* These large worms (up to eight or ten inches long) bear sharp bristles that give them a furry caterpillar-like appearance. The innocent-looking bristles, called setae, are actually hollow needles that inject a painful poison. They should not be handled with bare hands.

Some fishes are well known for their toxic stings. The most noted, of course, is the Indo-Pacific stonefish. These creatures have thirteen dorsal spines, each of which is equipped with a sac-like poison gland at its base. The spines are hollow, like hypodermic needles. The fish spends much of its time resting on the bottom, where it could be easily stepped on by a wader. If a person does step on the fish, the poison will be injected through the spine into his foot. The effect is said to be one of the worst pains known to man, and an encounter can be fatal in as short a time as fifteen minutes if the spine chances to enter a blood vessel directly. The stonefish is exceptionally well camouflaged in shape and color. It resembles an algae-covered rock and relies on this protective resemblance to the sea bottom, remaining motionless when approached. Divers and fishermen occasionally step on one without even seeing it.

Almost as well known is the lion fish. Although brilliantly colored, perhaps as a warning to potential predators, it nevertheless is quite well camouflaged to the human eye. Quite fearless, it often glides through the water with complete unconcern. Both the lion fish and the stonefish have poison glands associated with their dorsal-fin spines. They are both members of the scorpion-fish family, but lion-fish venom is not quite as virulent as the stonefish's, and its delivery system is not as well developed. This matters little, however, as the single prick of a lion-fish fin will cause an excruciating, incapacitating agony which can last for several hours. Fortunately not all members of the scorpion-fish family are equally dangerous. Some species are quite potent; others, such as the spotted scorpion fish, are scarcely toxic at all.

In near-shore coastal waters, fishermen often encounter sting rays, which lie partly buried in the sand. If one should step on a ray, the fish will defend itself by driving its double saw-edged dorsal spine into one's ankle. In addition to inflicting mechanical damage the spine has a groove running along its length that is filled with poison-producing cells; the resulting wound can be quite serious and, in fact, fatal.

Certain fishes use the spines on their gill covers for defense, and one group of toadfish even has a hollow opercular spine that functions like a hypodermic needle. Members of the surgeon-fish family have a pair of sharp spines just in front of the tail fin. When molested or in a territorial combat, the surgeon fish curves its tail so that a spine sticks out like a half-opened jackknife. Most of these fish merely cause mechanical injury, but at least one species is known to have a poison gland associated with the caudal spines. Other fishes such as sea catfish of the family Plotsidae have poisons associated with their dorsal and pectoral fin spines, and their stings can be quite serious.

Poison glands are generally modified skin-slime glands and show various degrees of elaboration. Most fish slimes are at least slightly toxic, and this is the reason that a puncture from a spine always causes more pain than an ordinary injury of the same severity. In fact, some fishes such as the trunk-fish and the cowfish have slime that is toxic enough to kill other fishes kept in the same container with them, although they lack special venomous spines.

Quite different from the venoms we have been talking about is the poison that causes sickness if the animal is eaten. Several groups of fishes are poisonous. Presumably this is an advantage to the species because a predator that eats one individual will become sick enough to learn to avoid the species the next time. Obviously this is conjecture, for if the predator dies, he won't learn to avoid that prey, and of course in either case it does the fish who is eaten no good to participate in the education of the predator. Moray eels, puffers, and a few other groups of fishes have some members that are toxic. Other members of the same families and even genera are not poisonous and may be desired food fishes. In the case of the puffers there are some species that produce mild hallucinations, and in Japan they are prepared under specially controlled conditions to preserve this effect. On the other hand, the puffers (blowfish) of our Atlantic coast are desirable commercial fish. Apparently they contain such minute amounts of poison that one cannot eat enough puffers at one time to suffer any ill effect from them.

A more difficult kind of fish poison to work with is that which produces a bad effect called ciguatera or ichthyosarcotoxism. This is due to substances that are taken in by the fish in its food and not to any property of the fish itself. For this reason a species that is normally safe to eat may suddenly come to be poisonous in a particular area. There has been a great deal of work on this subject because of its profound importance to the commercial fishing interests. One theory has it that the actual toxin is produced by plants, probably simple algae; the toxin is concentrated by the fishes that feed on the algae, then further concentrated at each link in the food chain until the large predaceous fishes such as snappers, barracudas, and jacks have enough concentrated toxin to affect human beings seriously. Symptoms of ciguatera include nausea, diarrhea, and nervous disturbances starting with general tingling sensations, and finally paralysis and death. In many cases local populations avoid large fish of certain species known to be poisonous. Unfortunately there is no way to test a fish in advance.

Finally there are in warm parts of the Indo–Pacific region and off the Pacific coast of Central America a few species of sea snakes that live in the ocean. For the most part they are fairly docile. They swim in the open ocean as well as near river mouths and around reefs. One can often find a sea snake poking its head between coral branches, looking for the small fishes on which it feeds. These snakes have a powerful neurotoxic venom with which they apparently subdue their prey. Sea snakes can be recognized by thin scales (most eels, with which they might be confused, have no scales) and their compressed, flattened tails. And of course, unlike eels, which are fishes, sea snakes must come to the surface to breathe. Because of their general docility, sea snakes have been thought of as not dangerous. While it is true that some species can barely be induced to attack, others are quite aggressive. Most bites have resulted from attempts to handle the snakes, and a large percentage of all known bites have been fatal.

While it is true that many marine organisms are able to defend themselves or at least retaliate if they are molested by a human intruder, it must be pointed out that the greatest danger to man comes not from the plants and animals but from other physical factors. Abrasions from sharp rocks are much more common than urchin-spine injuries, and the most imminent danger is always drowning as a result of fatigue or equipment failure. To a well-trained and cautious diver the dangerous marine animals are actually less dangerous than the mountain lions, rattlesnakes, and briar patches of the land. In the sea, as on land, the most dangerous animal is man.

O sweet spontaneous
earth how often have
the
doting

     fingers of
prurient philosophers pinched
and
poked

thee
, has the naughty thumb
of science prodded
thy

   beauty   , how
often have religions taken
thee upon their scraggy knees
squeezing and

buffeting thee that thou mightest conceive
gods
     (but
true

to the incomparable
couch of death thy
rhythmic
lover

    thou answerest

them only with

        spring)

             e.e. cummings

10. NOTES

These Notes are designed to give a more complete background of information than would normally be included in captions. The information is divided into two parts. The first part gives the common name of the animal or animals, followed in the same order by the scientific name or names. In a number of instances the animals had no specific local common names, so I have invented ones that seemed appropriate; this is the way animals come to be named anyway. The scientific names have been supplied when available. A question mark indicates that the animal could not be precisely identified from the photograph, or that it is a species that is so little known that it has not yet been given a formal designation. In some cases only the generic name is given; the details that divide one species from another are sometimes so minute that they cannot be determined except by much closer examination than is possible when one is photographing live animals in their natural surroundings. Following the scientific name is the approximate size of the animal as it is reproduced in the book, in relation to its actual dimensions. In presenting the photographs, I have tried to show the subjects life size or slightly larger, so that they are easy to see. However, since the animals were not collected and measured, the sizes quoted are approximate. The size is followed by the place where the photograph was taken. When no local or formal name for a particular location exists, a descriptive one has been supplied (such as east or west reef), along with the name of the adjacent district, town, island, or country. After the location is the approximate depth at which the photograph was taken. The pictures were taken anywhere from 1 foot below the surface to as much as 240 feet down, and since marine life varies depending on the depth, it is useful to know approximately where the animal was photographed and can be found. Last, I felt it important to designate the month and year that the photograph was taken, primarily for historical purposes. The photographs were taken over a period of seven years of intensive work.

In the second part of each note I have tried to give specific ideas or theories about life in the sea, drawn from my general background of experience gained either from reading or from personal observations while working under water. Although each individual note applies to a specific photograph, I have tried to write it in such a way that the notes can be read from beginning to end as a continuing story.

The small black and white photographs which accompany each note provide a pictorial index to the color plates, and follow the same order.

LIFE ON A PILING • ? • ½ life size • Hodgkins Cove, Gloucester, Massachusetts, U.S.A. • 10 feet • August 1967

Page 13

Marine animals live together in communities, as this picture of a section of a piling shows, and their lives are much affected by one another. Sponges, mussels, and anemones feed on plankton. The piling, which is subject to food-bearing currents, offers a good home for sedentary animals. Here sea urchins are feeding on minute organisms that also live on the piling, and the solitary starfish is consuming a mussel. This is a rich community but, by necessity, it is not overcrowded.

HAWKSBILL TURTLE • *Eretmochelys imbricata* • ½ life size • Barrier reef channel, Aulong Island, Palau Islands • 6 feet • October 1967

Pages 14-15

Life in the sea is exceedingly varied. The field of staghorn corals utilizes air dissolved in the water, while the free-swimming turtle, like all reptiles, must return to the surface for air. Unlike land turtles, sea turtles have evolved winglike legs for propelling themselves through their watery medium, as birds do in the air.

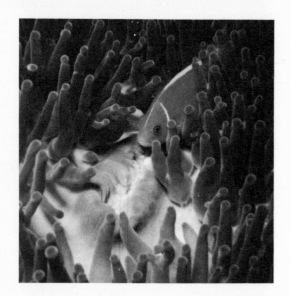

SKUNK CLOWNFISH AND ANEMONE • *Amphiprion perideraion* and *Radianthus* sp. • Twice life size • Point Lefévre reef, Lifou, Loyalty Islands • 60 feet • September 1965

Page 16

The relationship between clownfishes and their anemones varies considerably. The skunk clown actually eats from the mouth of its anemone, nuzzling in the mouth and removing either partially digested food or wastes. In a family of skunk clowns, there is a pecking system which preserves this source of food. When a juvenile skunk clown tries to steal a morsel of food, one of the adults usually chases the upstart away.

SARGASSUM FISH • *Histro histro* • 3 times life size • Public beach, Delray Beach, Florida, U.S.A. • 1 foot • June 1963

Page 17

Some animals stand out from their environment because, for various reasons, they need not be inconspicuous. The sargassum fish represents the opposite extreme. It has evolved into an indistinguishable part of the sargassum seaweed which is its total environment. The mottled markings and skin flaps enable it to blend into the scenery. The sargassum fish and many other protectively colored animals such as shrimps, pipefish, and sea slugs make their home in the floating algae.

FORMOSA JELLYFISH • *Olindioides formosa* • 1½ times life size • Tanabe Bay, Shirahama, Wakayama-ken, Japan • 50 feet • June 1966

Pages 18-19

Looking more like a gay party hat or a decorated flying saucer than a form of sea life, this jellyfish does not need to be inconspicuous, for it has powerful stinging cells for self-protection and for catching prey. The sea wasp, a not-too-distant relative, is one of the most deadly creatures in the sea. The sting from an adult can kill a man in a few minutes.

JELLYFISH • ? • 1¼ times life size • Jellyfish Salt Lake, Koror Island, Palau Islands • 3 feet • August 1969

Page 20

Not all jellyfish are dangerous stingers. This one has a sting that is very mild (at least for human beings), probably because it feeds primarily on small plankton. It has been said that jellyfish are not much more than organized water, for their body content is about 95 per cent water, 4 per cent salts, and 1 per cent organic matter.

PLUMOSE ANEMONE • *Metridium senile* • Life size • Hodgkins Cove, Gloucester, Massachusetts, U.S.A. • 12 feet • August 1967

Page 21

This is a common cold-water anemone that lives on pilings and rock walls in relatively protected waters. The delicate tentacles are extended into the current to trap plankton. In British Columbia waters I have observed them trap moon jellyfish that accidentally collided with the anemone. The jellyfish, being too large for immediate absorption, is held by the tentacles and rotated slowly until it is entirely digested.

SHORTHORN SCULPIN • *Myoxocephalus scorpius* • 1¼ times life size • Hodgkins Cove, Gloucester, Massachusetts, U.S.A. • 20 feet • August 1967

Page 22

Although science frowns upon anthropomorphic interpretations on the part of its investigators, the photographer is free to point out the likenesses to human beings that are present in other animals. Actually, this is the natural way man relates to his environment. Encountered head-on through the lens, the shorthorn sculpin has something endearingly human about it.

DEVIL SNAPPER • *Lutjanus* sp. • ½ life size • Jellyfish Salt Lake, Koror Island, Palau Islands • 15 feet • August 1969

Page 23

If the initial inclination to make an anthropomorphic judgment of absolute evil can be avoided, then one is free to appreciate the delicate structure of the pectoral fins of this snapper as they are extended to brake the fish's forward movement. The fish is able to see straight ahead, despite the fact that its eyes are situated on the sides of the head.

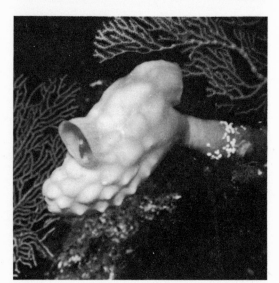

STALKED ASCIDIAN • *Polycarpa aurata* • 1½ times life size • Mouillage d'Amère, New Caledonia • 75 feet • September 1965

Page 24

Tunicates or sea squirts, because they are usually sedentary, can easily be mistaken for sponges, which they somewhat resemble in appearance and habits. Like sponges, sea squirts filter food, but they are much more highly evolved animals. The juveniles swim freely, like tiny tadpoles, until eventually they attach themselves to the bottom, head down and, in a sense, degenerate into sedentary adults.

FEATHER-DUSTER WORM • *Sabellastarte indica* • 2½ times life size • Tanabe Bay, Shirahama, Wakayama-ken, Japan • 35 feet • June 1966

Page 25

Tube worms, with their delicate gills, are perhaps the loveliest of the marine worms. They retract their feather-like gills with lightning speed when disturbed, and with good reason, for some blennies feed on the gills and are able to strike with equal speed.

OPAL SWEEPERS • *Parapriacanthus* sp. • 1½ times life size • Barrier reef channel, Aulong Island, Palau Islands • 15 feet • October 1967

Page 26

In the sea many fish may act and react as individuals, but when they are members of a school they function as parts of what seems like a single organism. The members maintain a definite space between one another, and the leaders are interchangeable, depending on immediate circumstances. If one individual in the school is attacked, his reaction triggers a response from the entire group. It is believed that the size and something about the general appearance of a school may serve to intimidate predators.

UDDER SPONGES • *Leucetta floridana* • Life size • The wall, Small Hope Bay, Andros Island, Bahamas • 220 feet • August 1968

Page 31

Although sponges are ranked low on the ladder of evolutionary complexity, it is important to realize that no organism is really simple. Life is extremely complex, and sponges have proved their ability to survive along with all the more sophisticated animals. There is always the possibility that sponges, being less specialized, will endure long after some of the more complicated animals become extinct.

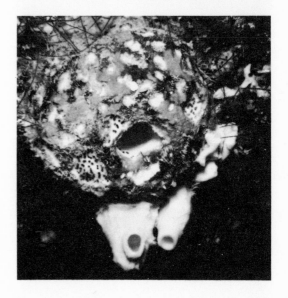

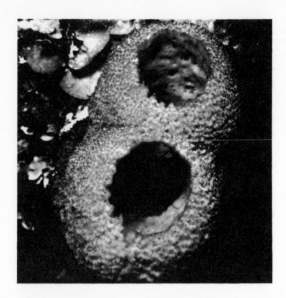

STRAWBERRY SPONGES • *Mycale* sp. • 1½ times life size • The wall, Small Hope Bay, Andros Island, Bahamas • 120 feet • July 1965

Page 32

The lovely rich red color of this sponge is lost 120 feet down in the undersea world, where many red animals appear to be dull brown or black. Its brilliance is revealed only by light from a flashbulb.

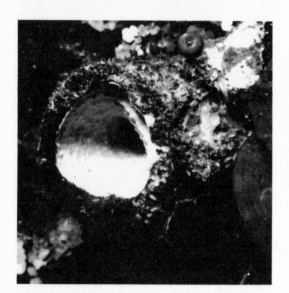

YELLOWMOUTH SPONGE • ? • Life size • The wall, Small Hope Bay, Andros Island, Bahamas • 100 feet • July 1965

Page 33

Unlike the reds, shades of yellow are still visible to the unaided eye at depths of 100 to 150 feet, because the water, which acts as a filter, takes longer to eliminate yellows. The most beautifully colored sponges usually occur in depths below 30 or 40 feet.

ORANGE COLONIAL SPONGES • ? • Life size • Cyclone Point reef, Heron Island, Queensland, Australia • 50 feet • October 1965

Page 34

This colony of sponges, like many others, grows under a ledge along with other encrusting organisms, including the small pink tube corals and a different species, the blue sponge. This photograph shows the tiny openings on the outer walls of the sponges through which the water is taken and filtered for minute organisms. The large central openings facilitate the return of the water to the environment.

VOLCANO SPONGES AND BRAIN SPONGES • *Ircinia* sp. and *Didiscus* sp. • ¾ life size • White Point outer reef, Bay Islands, Honduras • 60 feet • October 1964

Page 35

A rich community of invertebrate life can coexist in the small space of a few cubic feet of ocean environment. Here not only sponges but corals, gorgonians, bryozoans, and algae compete for living space.

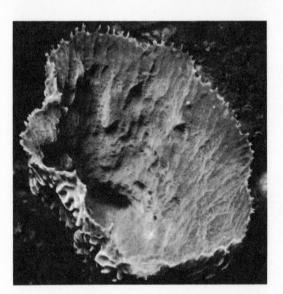

CHALICE SPONGE • *Callyspongia plicifera* • ¾ life size • White Point outer reef, Bay Islands, Honduras • 60 feet • September 1964

Page 36

In the sea there are many animals that fluoresce under ultraviolet light, and in the twilight depths some emit a soft glow. Normally the brightness of the flash destroys this effect, but once in a while a faint hint of the glow remains visible, as on the surfaces of this sponge.

POTTERY SPONGES • *Agelas* sp. • ¾ life size • The wall, Small Hope Bay, Andros Island, Bahamas • 110 feet • July 1965

Page 37

This colony of sponges probably started as a single sponge which grew and multiplied by splitting. The sponges in the lower foreground are in the process of doing just that.

FAN SPONGE • *Phakellia folium* • 1½ times life size • The wall, Small Hope Bay, Andros Island, Bahamas • 240 feet • August 1968

Page 38

Unlike the shallow surface waters, the deeps are relatively tranquil, so some of the animals that live in these quiet regions need not have a strong superstructure to exist. Many are quite fragile.

RED SEA WHIPS • ? • 1½ times life size • Barrier reef channel, Aulong Island, Palau Islands • 40 feet • October 1967

Page 43

Corals offer one of the sea's richest areas for photography. Their forms and colors are a delight to the eye. This gorgonian's natural structure allows each polyp living space. Here the tiny polyps are expanded and are feeding on the plankton riding the incoming tide.

TREE CORALS • ? • Life size • The Creek, Red Sea, Obhor Kuraa, Saudi Arabia • 40 feet • March 1965

Page 44

The alcyonacean tree corals, unlike the gorgonians and stony corals, collapse into blobs when they are not feeding. This photograph shows the corals erect and feeding. The tiny spicules on the trunks and limbs of the corals give them something like an elastic rigidity. However, this again is an example of an animal relying on the water for primary structural support.

STINGING BUBBLE CORAL • ? • 1½ times life size • Ngadarak reef, Malakal Harbor, Urukthapel Island, Palau Islands • 40 feet • November 1967

Page 45

This coral, with its polyps contracted for the moment, displays an intricate skeletal structure that somewhat resembles a maze. However, unlike the bleached white "corals" for sale in curio shops, this skeleton is covered with the delicate hues of living flesh.

SAUSAGE CORAL • ? • 1½ times life size • Ngadarak reef, Malakal Harbor, Urukthapel Island, Palau Islands • 100 feet • November 1967

Page 46

Usually the fluorescent corals are found in deeper waters or under ledges. Unlike some of the others, fluorescent corals are mostly active during the daylight hours.

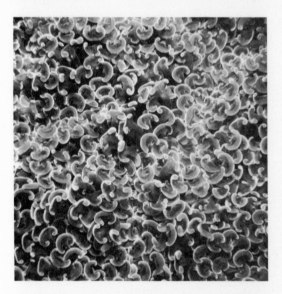

DAISY CORAL • *Goniopora lobata* • 1½ times life size • Ngadarak reef, Malakal Harbor, Urukthapel Island, Palau Islands • 45 feet • October 1967

Page 47

Looking more like flowers or stars than like corals, these delicate creatures expand their polyps from a common stony skeleton. The petal-like tentacles wait patiently for the tide-borne food. Even among scientists it is commonly thought that most corals feed only at night. Actually, many factors influence their feeding, one of the most important being the tides. Many corals are active primarily when water is moving, and are inactive at slack tide. All these photographs of corals were taken during daylight hours, and all but one show them feeding.

SOLITARY FLUORESCENT CORAL • *Euphyllia picteti* • Life size • Banc Gail, New Caledonia • 120 feet • September 1965

Page 48

Most coral polyps are very small, but a number of solitary corals, such as the mushroom corals, have a single polyp that is quite large, with calyxes eight to ten inches in diameter. Many fluorescent corals, like the one illustrated, are solitary and not colonial. Under ultraviolet light this one glows a lovely chartreuse, and the tentacles are tipped with red.

GREEN TUBE CORAL • *Dendrophyllia* sp. • Life size • Barrier reef channel, Aulong Island, Palau Islands • 15 feet • October 1967

Page 49

Most of the polyps of this green tube coral are reaching out to feed, but there are some that are only partially expanded, and a few that are totally contracted. Each individual polyp is able to have some semblance of an individual existence, although there is probably no exact point where one polyp begins or ends. Each is joined to those around it by the common flesh covering the calcium skeleton.

ORANGE TUBE CORAL • *Dendrophyllia arbuscula* • 2¼ times life size • Tanabe Bay, Shirahama, Wakayama-ken, Japan • 50 feet • June 1966

Page 50

Japan's waters are not particularly cold, but the temperature is low enough to prevent the existence of reef-forming corals. Although corals are found in all waters, cold-water corals are small solitary formations, and like this lovely specimen, they are often found growing under ledges. Note that the lower right coral polyp is touching a tentacle to its mouth. The delicate yellow tentacles trap food and transfer it in this way.

CHAMBERED NAUTILUS • *Nautilus macromphalus* • Life size • Mbere Reef, New Caledonia • 100 feet • September 1965

Page 55

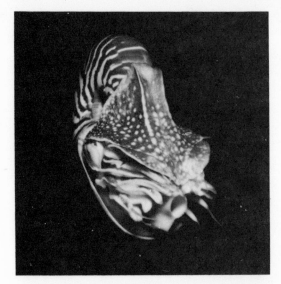

The chambered nautilus, a living fossil, has always been a fascinating creature. The animal is a cephalopod, but, unlike the octopus or cuttlefish, it has very primitive eyes that function much like a pinhole camera. Its siphon tube is only partly formed and therefore does not allow the nautilus to jet-propel itself through the water as quickly as a squid. Nautiluses usually live in very deep water down to several thousand feet during the daytime, but at night they may come up to within one hundred feet from the surface, to feed.

CUTTLEFISH • *Solitosepia liliana* • ½ life size • Amédée Island reef, New Caledonia • 8 feet • September 1965

Pages 56-57

In spring—September in the Southern Hemisphere—many animals mate and lay their eggs. This rare photograph shows two cuttlefish mating. The male, which is on the right, first approached the female and signaled to her by flashing blue colors over his body. As he approached closer, he raised his arms and stroked her forehead. She in turn raised her arms to receive him, and they embraced. While mating, they glided over the reef. After a few minutes they parted, possibly never to meet again.

OCTOPUS • *Octopus* sp. • Life size • Tanabe Bay, Shirahama, Wakayama-ken, Japan • 30 feet • June 1966

Page 58

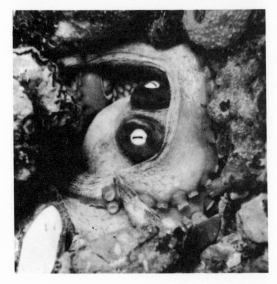

The octopus has only recently emerged from its role as the demoniacal creature of myth to the status of a relatively gentle, timid animal which tries to avoid encounters with a much more dangerous creature, man. This octopus is shown in its burrow with the leftovers of a distant relative, a bivalve that it ate for dinner. Octopuses have powerful parrot-like beaks for subduing prey, and some have accompanying venom glands. In Australia, one practical joker slapped a tiny species of octopus on a friend's bare back. The bite was like a pin-prick, but two hours later the man was dead.

ROOSTER-COMB OYSTER • *Lopha cristagalli* • 1½ times life size • Kuapesngas Point reef, Malakal Harbor, Urukthapel Island, Palau Islands • 45 feet • September 1969

Page 59

Plankton feeders both, the oysters and the gorgonian to which they are attached are sharing a common niche where they are nourished by a current of water. The oysters take the water in through the small openings and filter it for food. The apparent red coloring of the oyster is actually an encrusting sponge which is also a filter-feeder.

THORNY OYSTER • *Spondylus* sp. • Life size • Mbere Reef, New Caledonia • 75 feet • September 1965

Page 60

The thorny oyster is actually a scallop. Living scallops, with their lovely mantles, are much more beautiful than many of the mollusks cherished by shell-collectors. Most mollusk shells are encrusted and must be cleaned before the beauty of the shell is revealed. Bivalves are particularly susceptible to encrustation by other colorful animals and plants, which adds to the beauty of the living animal while it is in its natural environment.

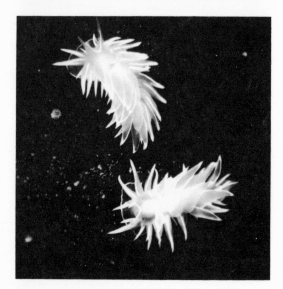

ALABASTER NUDIBRANCHS • *Dirona albaliniata* • 1⅓ times life size • West point, Brandon Island, Departure Bay, Vancouver Island, V.C., Canada • 35 feet • November 1969

Page 61

"Nudibranch" means "naked gill," and these creatures are all the more lovely because of this characteristic. They are shown here crawling slowly across a devil's-apron alga leaf, feeding on algae and other organisms.

CRIMSON NUDIBRANCH • *Ceratosoma cornigerum* • 1¼ times life size • Tanabe Bay, Shirahama, Wakayama-ken, Japan • 30 feet • June 1966

Page 62

Many nudibranchs are extremely beautiful and conspicuously colored, sometimes as a warning to would-be predators. Their flesh has an unpleasant taste, so they have no need of the protective shell characteristic of most mollusks.

LOBSTER • *Homarus americanus* • Life size • Hodgkins Cove, Gloucester, Massachusetts, U.S.A. • 20 feet • August 1967

Page 67

In the sea, there is no respect for near relatives. This lobster would evidently just as soon eat another crustacean as it would any other animal. The powerful claws aid in defense and allow it to crack open other shelled animals, including smaller lobsters.

HERMIT CRAB • *Pagurus* sp. • 3 times life size • Benjamin River, Sedgwick, Maine, U.S.A. • 30 feet • August 1967

Pages 68-69

It is thought that some hermit crabs relish the tube feet of sea urchins, and it is obvious here that this one is interested in something that belongs to the urchin.

BULLDOZER SHRIMP AND LOOK-OUT GOBY • Alpheus djiboutensis and Cryptocentrus sp. • Twice life size • Ngadarak reef, Malakal Harbor, Urukthapel Island, Palau Islands • 20 feet • September 1969

Page 70

This odd couple lives together in a hole which the shrimp builds and keeps free of debris by continuously plowing out the sand. While the shrimp is at work, the little goby stands guard at the entrance. Occasionally it grabs a mouthful of sand and filters it through the gills for tiny organic matter. Very often a pair of gobies and a pair of shrimps inhabit the same hole.

BLUE CRAB AND RAGGED SEA HARE • Callinectes sapidus and *Bursatella plei* • Life size • Port Largo, Upper Key Largo, Florida, U.S.A. • 18 Feet • April 1969

Page 71

This crab relishes the internal parts of the sea hare, which it is extracting from its dying victim. The crab, with its greater mobility and strength, looks to the more defenseless creatures for its food supply. The purplish ink rising like smoke from the sea hare is a substance emitted for possible defense, in this case to no avail.

JONAH CRABS • Cancer borealis • 1½ times life size • Hodgkins Cove, Gloucester, Massachusetts, U.S.A. • 20 feet • August 1967

Pages 72-73

The presence of the male Jonah crab, seen in place above the female, protects her while her new shell hardens. She has just molted, and her empty shell is lying upside down in front of them. As their shells do not grow, all crustaceans must molt periodically. The soft new shells make them quite defenseless. Jonah crabs mate shortly after the female molts, and the survival of the species is thus doubly served.

PEPPERMINT CLEANER SHRIMP AND FAIRY BASS •
Hippolysmata grabhami and *Anthias squamipinnis* •
1½ times life size • Nature Reserve, Gulf of Aqaba,
Eilat, Israel • 20 feet • April 1965

Page 74

This shrimp has established its cleaning station on a
small promontory under a ledge. A number of orange
fairy bass await their turn while the doctor is working
on one patient. Deftly he removes (and eats) parasitic
copepods from the body of the fairy bass. The copepods
are tiny crustaceans. Some cleaner shrimps are not as
well adapted to this work as others, and as a consequence are less immune from predation. Occasionally
the doctor ends up in the patient's stomach.

SEA STAR AND GREEN SEA URCHINS • *Asterias
vulgaris* and *Strongylocentrotus droehbachiensis* •
1⅓ times life size • Hodgkins Cove, Gloucester, Massachusetts, U.S.A. • 15 feet • August 1967

Page 79

The starfish makes good use of its water vascular system
when feeding on bivalves. By gripping both sides of the
mussel and then creating a vacuum in its tube feet, this
starfish is able slowly to overcome the mussel's power
to keep its shell closed. When the mussel tires and its
shell opens slightly, the starfish extends its stomach
around the soft parts of the mussel, and digestion begins. Possibly some of the green sea urchins will partake
of the feast.

SLATE-PENCIL URCHIN • *Hetercentrotus trigonarius* •
1½ times life size • Amédée Island reef, New Caledonia • 5 feet • August 1965

Pages 80-81

This sea urchin is atypical in that its spines are thick
and blunt, and its test is also armored with short stubby
spines. Like many of the tropical urchins, this species
is primarily active at night. During the day it wedges
itself under ledges or between branches of coral,
rather than relying totally on its own armor for protection against predators.

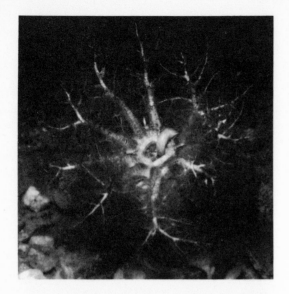

SEA CUCUMBER • Cucumaria frondosa • 1¼ times life size • Benjamin River, Sedgwick, Maine, U.S.A. • 30 feet • August 1967

Page 82

Many tropical sea cucumbers are not as obviously active as this cold-water species. One need only observe it for several minutes when the tide is running. The bushy arms collect plankton from the current, and with clocklike precision one arm after another is stuffed into the sea cucumber's central mouth, where it is gleaned of its catch. Approximately every thirty seconds there is a changing of the guard: one arm is removed and the next inserted.

SHORT-SPINED SEA URCHIN • Tripneustes gratilla • 1⅓ times life size • Tanabe Bay, Shirahama, Wakayama-ken, Japan • 25 feet • June 1966

Page 83

The tiny feet of this lovely urchin are visible just below the spines near the animal's leading edge. The urchin uses its feet to traverse the sea bottom. The spines serve to protect the urchin, but not always successfully. It is known that queen trigger fish, and no doubt other creatures, attack urchins despite their spines.

BLACK-AND-WHITE CRINOID • ? • Life size • Point Merlet reef, Lifou, Loyalty Islands • 90 feet • September 1965

Page 84

Some of the loveliest of the sea's creatures are the crinoids. These living fossils display every shade and color of the spectrum. Since they are plankton-feeders, they are often seen attached to gorgonians and sea fans that grow where the water currents provide ideal feeding conditions.

FEATHER CRINOID • ? • Life size • Nature Reserve, Gulf of Aqaba, Eilat, Israel • 20 feet • April 1965

Page 85

Many crinoids are out in the open day and night and may or may not be active, depending on the tides. Others hide under ledges during the day. At dusk they creep out onto the top of coral formations to feed. Some crinoids can either move slowly over the bottom or, if they choose, actually swim through the water by undulating their feathery arms, like ballet dancers.

BASKET SEA STAR • *Euryale* sp. • ⅓ life size • Nature Reserve, Gulf of Aqaba, Eilat, Israel • 20 feet • April 1965

Page 86

A basket sea star is active and feeds always at night. During the daytime it is curled up under a ledge, a mass of grotesque arms. At night it emerges from its cave, like some primeval sea monster in the immense darkness. It crawls slowly to the top of a coral formation and then miraculously, intricately transforms itself into a fragile work of art.

BROWN CLOWNFISH AND SEA ANEMONE • *Amphiprion bicinctus* and ? • Twice life size • The Creek, Red Sea, Obhor Kuraa, Saudi Arabia • 50 feet • March 1965

Page 91

Scientists are still not certain just what mechanism prevents an anemone from releasing its stinging cells and killing its clownfish. However, it is known that when a clownfish is separated from its anemone for several months and then reintroduced, the fish has to reacclimate itself by touching parts of its body to the anemone before it is able to swim freely among the tentacles. Consequently it is thought that the fish acquires an identifying slime from the anemone, and that fish unmarked by slime are killed and eaten if they inadvertently swim into the anemone.

GOLDEN BOXFISH • *Ostracion tuberculatus* • 1¼ times life size • Cyclone Point reef, Heron Island, Queensland, Australia • 35 feet • October 1965

Page 92

This charming little fish grazes undisturbed along the bottom because it is encased in a protective bony skeleton covered by only a thin layer of flesh. Some species emit a poisonous slime, which also helps to discourage would-be predators.

LONGSPINE SQUIRREL FISH • *Holocentrus rufus* • Life size • Hog Island, Bay Islands, Honduras • 3 feet • September 1964

Page 93

When this species of squirrel fish is alarmed, it produces a rapid grunting sound, and, as in response to the wave of a magic wand, a shark or barracuda that has heard the noise will rapidly appear. For this reason, it is wise to avoid alarming fishes that get excited and send out distress signals.

GOLDEN LONGNOSE BUTTERFLY FISH • *Chelmon rostratus* • 1¼ times life size • Cyclone Point reef, Heron Island, Queensland, Australia • 30 feet • October 1965

Page 94

Like many of the butterfly fishes, this one has a false-eye spot near its tail, a characteristic that tends to mislead its enemies. The predator, anticipating the forward movement of its victim in the same way that a hunter leads a bird in flight, ordinarily strikes at the head. More than once I have seen a butterfly fish with a gouge near the eye spot, showing that the predator has struck at the tail instead, inadvertently giving the butterfly fish a better opportunity to escape.

HARLEQUIN TUSK WRASSE • Lienardella fasciatus •
Life size • Cyclone Point reef, Heron Island, Queensland, Australia • 45 feet • October 1965

Page 95

It is not known whether this particular fish is territorial, but many reef fishes that are brightly colored and conspicuously marked are definitely territorial. The "Poster" markings offer immediate recognition to other fish of the same species. If one fish is occupying an area of reef and another of the same species recognizes its presence by means of the distinctive markings, the intruder will usually steer clear of the occupied territory.

ORANGE-SPOTTED FILEFISH • Oxymonacanthus longirostris • 1¼ times life size • Turtle Bay, Watamu, Kenya • 10 feet • October 1966

Page 96

A pair of filefish glides over a small staghorn coral head surrounded by algae. Like many of the diminutive reef fishes, these files are grazers and have tiny mouths at the end of longish snouts to facilitate feeding in crevices.

CARDINAL FISH • Archamia fucata • 1¼ times life size • The Creek, Red Sea, Obhor Kuraa, Saudi Arabia • 50 feet • March 1965

Page 97

These cardinal fish take partial shelter among the branches of a gorgonian. Cardinal fish often school among the branches of staghorn or other corals. However, security is not guaranteed. On a number of occasions I have seen a larger fish dart in with lightning speed and catch a victim.

143

*ROYAL GRAMMA • Gramma loreto • Life size •
White Point outer reef, Bay Islands, Honduras • 40
feet • September 1964*

Page 98

Usually royal grammas are seen swimming upside down
under ledges. When disturbed, they dart for their holes
in the coral rock. Since water produces a state of near-
weightlessness, a fish can swim right side up or upside
down. Fishes that live under ledges usually orient themselves to the bottom, and if the bottom happens to be
the roof of a cave, it matters little. In this instance the
gramma happens to be swimming as we think a fish
should—yet it is still somewhat like an astronaut on a
space walk.

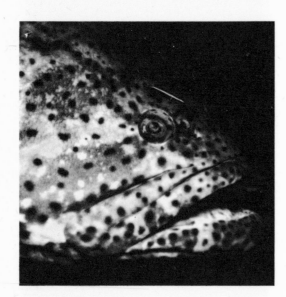

*NEON GOBY AND SPOTTED SEA BASS • Elacatinus
oceanops and Epinephelus itajara • ¾ life size • Patch
reef, Lower Matecumbe Key, Florida, U.S.A.   15 feet
• July 1965*

Page 103

The tiny 2-inch-long goby is one of the major Caribbean
cleaners. Usually a pair establishes a cleaning station
on a coral head or under a ledge. Even large fishes, such
as this 80-pound sea bass, come to the goby's cleaning
station to have parasites removed.

*NEON GOBIES, SPANISH HOGFISH, AND NASSAU
GROUPER • Gobiosoma genie, Bodianus rufus, and
Epinephelus striatus • ¾ life size • The Barge, Small
Hope Bay, Andros Island, Bahamas • 70 feet • August
1968*

Page 104

Three gobies and a juvenile Spanish hogfish go over a
grouper for parasites. The gobies are fairly full-time
doctors, but only the young of the Spanish hogfish act
as cleaners. This is true among other fishes that inhabit
Caribbean waters, such as French angelfish and blue-
head wrasses. They may dine on parasites while young,
but as they grow older their feeding habits change.

CLEANER WRASSE AND BLUE-SPOTTED SEA BASS
• *Labroides dimidiatus* and *Plectropomus maculatus* •
Life size • Cyclone Point reef, Heron Island, Queensland, Australia • 35 feet • October 1965

Page 105

There are at least four labrid cleaners in the Indo-Pacific tropical waters that act as the major doctor fishes for the rest of the fish population. The *Labroides dimidiatus* is by far the most abundant and widespread of the labrid wrasses. It readily establishes a cleaning station and is always eager to perform its services even on human beings, whose interest it mistakes for a willingness to be cleaned. This 4-inch adult wrasse will fearlessly enter the sea bass's mouth to pick parasites. Like many fishes, the sea bass extends its gills to facilitate cleaning. The gill rakers are easily visible in this picture. The wrasse may exit through the gills after completing its activities in the patient's mouth.

CLEANER WRASSES AND PINK FAIRY BASS • *Labroides rubrolabiatus* and *Mirolabrichthys tuka* • 1¼ times life size • Tuanoa Pass, Tahiti, Society Islands • 25 feet • December 1962

Page 106

These cleaners have a less well-defined station than *L. dimidiatus,* but they are full-time cleaners and will range over several square yards of reef area looking for patients.

CLEANER WRASSE AND SCORPION FISH • *Labroides bicolor* and ? • 1¼ times life size • Tuanoa Pass, Tahiti, Society Islands • 40 feet • November 1962

Page 107

Unlike the *L. dimidiatus,* which runs a clinic, the adult *L. bicolor* is a doctor on continuous house call. The juveniles start off their cleaning activities under a single ledge, where they are visited by patients. At mid-growth, as in this picture, the pre-adult comes out of hiding and performs his activities in much the same manner as an adult *L. dimidiatus,* keeping to a fairly local cleaning station. Then, as a full-grown adult, it begins house calls. An area of 50 to 100 feet in any direction is covered. Large schools of fish, such as jacks, will be serviced as they swim in a circle. It is believed that open-ocean fishes come onto the reef specifically to be cleaned.

CLEANER WRASSE AND GOATFISH • *Labroides dimidiatus* and *Parupeneus* sp. • ¾ life size • Barrier reef, Bulari Pass, New Caledonia • 50 feet • February 1963

Pages 108-109

This is a perfect example of the great importance of cleaning to the fish population. The wrasse has a full waiting room while one goatfish is being doctored. Note the extended barbels and erect fins on the fish being cleaned, compared to the folded fins and barbels of the waiting patients. When the doctor approaches, the patient erects its fins to signal its desire to be cleaned and to facilitate the operation.

CLEANER WRASSE AND SQUIRREL FISH • *Labroides dimidiatus* and *Holocentrus sammara* • ¾ life size • Amédée Island reef, New Caledonia • 10 feet • January 1963

Page 110

This cleaner circled around the group of patients with a characteristic prancing motion which suggests both approach and flight, but as soon as one of the squirrel fish opened its mouth and gill covers, the little doctor immediately approached and began cleaning activities.

YELLOW-LIP SEA SNAKE • *Laticauda colubrina* • Life size • Amédée Island Reef, New Caledonia • 3 feet • August 1965

Page 115

For those accustomed to seeing snakes only on land, it seems strange to watch a snake swimming just above a mass of coral. It seems even stranger to descend to a reef 150 feet down and see one of them ascending from the coral toward the surface. All sea snakes are deadly poisonous. Some are aggressive, and others, such as this species, are relatively docile but nonetheless dangerous. Most accidents occur because people who think sea snakes won't bite are foolish enough to handle them without proper precautions.

SCORPION FISH • ? • 1½ times life size • Nature Reserve, Gulf of Aqaba, Eilat, Israel • 20 feet • April 1965

Pages 116-117

Although these two scorpion fish are protected with poisonous spines, they still can get into a very serious territorial fight. In this rare photograph, the fish on the left, in a final act of aggression, is biting the other on the head. In leading up to this, the two fish faced off less than an inch apart. Each blew sand in the other's face until one had had enough bluffing and attacked. No damage was done, however, as the loser was finally released and promptly departed.

BANDIT PUFFER • *Arothron nigropuntatus* • Life size • The Creek, Red Sea, Obhor Kuraa, Saudi Arabia • 40 feet • March 1965

Page 118

Puffers get their name from their ability to inflate their bodies with water when molested. The fish can be deadly poisonous when eaten. The skin, liver, gonads, and intestines contain a powerful nerve poison. Yet in Japan puffer meat is relished as a delicacy. Some restaurants have specially trained *fugu* cooks to prepare the edible parts of the fish properly. Still, self-styled *fugu* cooks do not always live long lives.

LION FISH • *Pterois radiata* • 1¼ times life size • The Creek, Red Sea, Obhor Kuraa, Saudi Arabia • 60 feet • March 1965

Page 119

Poised upside down under a ledge, this lion fish displays its finery of fins and poisonous spines. Its unmistakable markings warn would-be predators to steer clear. Lion fish move slowly and are not easily alarmed. They probably know from experience that their enemies will back off fast after encountering a spine or two. The poison is more deadly than cobra venom and excruciatingly painful, even if one is lucky enough to survive a puncture wound.

STONEFISH • *Synanceja verrucosa* • Life size • Vata Bay, Noumea, New Caledonia • 5 feet • August 1965

Page 120

The stonefish is apparently the most venomous fish in the ocean. It has thirteen dorsal spines, three anal spines, and two pelvic spines, and all are equipped with venom glands. Pressure on the end of the spine will cause it to penetrate the source of the irritation and at the same time inject venom in an action like that of a hypodermic needle. Sometimes waders in shallow water accidentally step on the erected spines, and the pain is often so great that the victim screams and thrashes wildly about before losing consciousness. In extreme cases, the sting can be fatal.

CROWN-OF-THORNS SEA STAR • *Acanthaster planci* • ²/₃ life size • Amédée Island reef, New Caledonia • 15 feet • September 1965

Page 121

Looking like an exquisite sunburst, this starfish crawls over a mass of staghorn corals. Its beauty masks a double menace. The spines are venomous and produce painful puncture wounds when touched. In addition, the animal eats coral, and at present there is an epidemic of these starfish in the tropical Pacific. Something has disturbed the natural balance of nature, and the crown-of-thorns has been multiplying swiftly. Large areas of the Great Barrier Reef, as well as other reefs, have been destroyed. It may take centuries for new corals to repopulate the devastated areas.

BULL SHARK • *Carcharhinus leucas* • Actual length 7 feet • Wax Cut reef, Andros Island, Bahamas • 60 feet • August 1968

Page 122

When one fears sharks and knows they are as formidable as man and sometimes as ruthless, it is difficult to appreciate their graceful beauty. Yet they move with a silent speed through their blue world, acting out a perpetual ballet of calm and violence, life and death.

148

ACKNOWLEDGMENTS

The photographs for this book would not exist without the help of:
Caesar Antoniacci, Chūichi Araga, Edward Barnard, George Benjamin, Morten Beyer, Duane Busch, Ida Catala-Stucki, Drayton Cochran, John Cochran, David Dickerson, Robert Endean, Louis Eschembrenner, Sally Faulkner, Seymour and Alice Faulkner, Herbert Forgash, Carl Gage, Thomas Gibson, Charles and Julianne Golding, Ward and Maryke Griffioen, Bruce Halstead, Patricia Hunt, Leon and Audrey Israel, Finn Jensen, Leonard Jossel, John Kochi, Yves and Germaine Merlet, David Miller, Charles Monin, Phyllis Montgomery, Donald Moyer, Sylvain Napoleon, Robert and Hera Owen, Toshiro Paulis, John Porter, Reg and Shirley Rice, Huzio Utinomi, Richard Vahan, Anders Wästfelt, Peter Wilson, and Robert Woodward.

The authors also wish to thank for their help in preparing this book: James Atz, Frederick Bayer, Nicolas Ducrot, William Emerson, Barraclough Fell, Willard Hartman, Bryan Holme, Ludmila Karameros, Mary Kopecky, Ernest Lachner, Carlo Mosca, Leonard Nones, William Old, John Randall, Richard Randall, Allen Vogel, and Irwin Wolf.